ARSENIC IN RICE

A Global Menace

ARSENIC IN RICE

A Global Menace

Edited by

Garima Awasthi, PhD

Sudhakar Srivastava, PhD

Mahipal Singh Sankhla, MSc

AAP | APPLE ACADEMIC PRESS

First edition published 2024

Apple Academic Press Inc.
1265 Goldenrod Circle, NE,
Palm Bay, FL 32905 USA

760 Laurentian Drive, Unit 19,
Burlington, ON L7N 0A4, CANADA

CRC Press
2385 NW Executive Center Drive,
Suite 320, Boca Raton FL 33431

4 Park Square, Milton Park,
Abingdon, Oxon, OX14 4RN UK

Apple Academic Press exclusively co-publishes with CRC Press, an imprint of Taylor & Francis Group, LLC

Library and Archives Canada Cataloguing in Publication

Title: Arsenic in rice : a global menace / edited by Garima Awasthi, PhD, Sudhakar Shrivastava, PhD, Mahipal Singh Sankhla, PhD.
Names: Awasthi, Garima, editor. | Shrivastava, Sudhakar, editor. | Sankhla, Mahipal Singh, editor.
Description: First edition. | Includes bibliographical references and index.
Identifiers: Canadiana (print) 20230513700 | Canadiana (ebook) 20230513794 | ISBN 9781774914663 (hardcover) | ISBN 9781774914670 (softcover) | ISBN 9781003408390 (ebook)
Subjects: LCSH: Rice—Effect of arsenic on. | LCSH: Soils—Arsenic content. | LCSH: Groundwater—Arsenic content.
Classification: LCC SB191.R5 A77 2024 | DDC 633.1/899—dc23

Library of Congress Cataloging-in-Publication Data

CIP data on file with US Library of Congress

ISBN: 978-1-77491-466-3 (hbk)
ISBN: 978-1-77491-467-0 (pbk)
ISBN: 978-1-00340-839-0 (ebk)

About the Editors

Garima Awasthi, PhD

Assistant Professor, Vivekananda Global University, Jaipur, Rajasthan, India

Garima Awasthi, PhD, is an Assistant Professor at Vivekananda Global University, Jaipur, Rajasthan. She earned her PhD in Biological Sciences from CSIR-National Botanical Research Institute (CSIR-NBRI) in 2016. She also worked as a SERB-National Post Doc Fellow at the University of Lucknow from 2016–2018. In recent years, she has published more than 13 scientific articles in internationally indexed and peer-reviewed journals on heavy metal stress in plants, mainly on the arsenic problem in rice, which has been widely cited. She has been an evaluator of scientific articles and manuscripts for various international scientific publishers. The line of research that she develops is multidisciplinary in the areas of plant abiotic and biotic stress, nanotechnology, plant physiology, and proteomics. She has collaborated with various educational institutions and research centers, as well as with the industrial sector. Dr. Awasthi has taught various courses and seminars and has participated in national and international scientific conferences. Likewise, she has advised various students for the development and completion of bachelor\'s degrees, Master of Science degrees, and Doctor of Science theses in the life sciences discipline.

Sudhakar Srivastava, PhD

Assistant Professor, Institute of Environment and Sustainable Development, Banaras Hindu University, Varanasi, Uttar Pradesh, India

Sudhakar Srivastava, PhD, is an Assistant Professor at the Institute of Environment and Sustainable Development (IESD), Banaras Hindu University (BHU). He earned a PhD in Botany from the

CSIR-National Botanical Research Institute (CSIR-NBRI) and the University of Lucknow in 2008. He served as Scientific Officer at Bhabha Atomic Research Center (BARC), Mumbai, from 2009 to 2014 before joining BHU. He has more than 19 years of research experience in the field of plant-metal interactions. He has published more than 125 research and review articles in high-impact international and national journals. He has also published 25 book chapters and edited three books. He has supervised several PhD and several MPhil students. Dr. Srivastava an associate of the National Academy of Agricultural Sciences. He received a Young Scientist Award from the National Academy of Sciences, India (NASI), Allahabad, in 2011 and Young Scientist Award from the Uttar Pradesh Council of Science and Technology (UPCST) in 2013–2014. Dr. Srivastava is also a Fellow of the Indian Society of Environmental Botanists (ISEB), Lucknow.

Mahipal Singh Sankhla, MSc

Assistant Professor, Department of Forensic Science, University Center for Research and Development (UCRD), Chandigarh University, Mohali, Punjab, India

Mahipal Singh Sankhla, MSc, is a highly accomplished Assistant Professor in the Department of Forensic Science, University Center for Research and Development (UCRD), Chandigarh University, Mohali, Punjab, India. With extensive experience in the field of forensic science, he previously served as an Assistant Professor in the Department of Forensic Science at Vivekananda Global University in Environmental Science and Pollution Research, Jaipur, Rajasthan. Mr. Sankhla earned a Bachelor of Science (Hons.) degree in Forensic Science and a Master of Science degree in Forensic Science. He is currently pursuing his PhD in Forensic Science at Galgotias University in Greater Noida, U.P. Mr. Sankhla has received training from several renowned laboratories, including the Forensic Science Laboratory (FSL), Lucknow; CBI (CFSL) New Delhi; Codon Institute of Biotechnology Noida; and Rajasthan State Mines and Minerals Limited (R&D Division) Udaipur. He has received several awards and accolades for his outstanding work, including a Junior Research Fellowship (JRF) from the DST-funded project at Malaviya National Institute of Technology—MNIT, Jaipur; a Young Scientists Award, Young Researcher Award, and a Forensic Researcher Award. He has also been credited with one Indian copyright and one German patent. Mr. Sankhla has edited five books and published over 30 book chapters with various national and international publishers. His research work has been published in more than 160 peer-reviewed national and international journals. His primary research areas include heavy metals toxicity, forensic toxicology, environmental toxicology, nanotechnology, and fingerprint analysis.

Contents

Contributors

Rakhi Agarwal
School of Forensic Sciences, National Forensic Sciences University, India

Ankita
Department of Anthropology, Panjab University, India

Vinay Aseri
Department of Forensic Science, Vivekananda Global University, India

Anjali Awasthi
Department of Zoology, University of Rajasthan, India

Garima Awasthi
Department of Life Sciences, Vivekananda Global University, India

Kumud Kant Awasthi
Department of Life Sciences, Vivekananda Global University, India

Tejasvi Bhatia
Department of Forensic Science, School of Bioengineering and Biosciences, Lovely Professional University, India

Jayanta Kumar Biswas
Department of Ecological Studies and International Center for Ecological Engineering, University of Kalyani, India

Rushikesh L. Chopade
Department of Forensic Science, Vivekananda Global University, India

Mahesh Das
Department of Botany, Gauhati University, India

Payal M. Deoghare
School of Medico-Legal Studies, National Forensic Sciences University, India

Sonali Dubey
Institute of Biotechnology, Amity University, Noida, Uttar Pradesh, India

Sanjay Dwivedi
Council of Scientific and Industrial Research–National Botanical Research Institute (CSIR-NBRI), Lucknow, Uttar Pradesh, India; Academy of Scientific and Innovative Research (AcSIR), Ghaziabad, Uttar Pradesh, India

Nandini Gupta
School of Forensic Sciences, National Forensic Sciences University, India

Sunil K. Gupta
Key Laboratory of Tropical Forest Ecology, Xishuangbanna Tropical Botanical Garden, Chinese Academy of Sciences, China

Sabnoor Yeasrin Jyoti
Department of Botany, Gauhati University, India

Bikash Kalita
Department of Botany, Gauhati University, India

Indrajit Kalita
Department of Botany, Gauhati University, India

Jyotirmay Kalita
Department of Botany, North Gauhati College, India

Navjot Kaur
Department of Forensic Science, Chandigarh University, Mohali, Punjab, India

Anshika Kaushik
Department of Anthropology, University of Delhi, India

Rubina Khanam
ICAR–National Rice Research Institute, India

Dipak Kumar Mahida
Department of Biochemistry and Forensic Science, Gujarat University, India

Meghna
Department of Forensic Science, School of Bioengineering and Biosciences, Lovely Professional University, India

Amarnath Mishra
Amity Institute of Forensic Sciences, Amity University, India

Varad Nagar
Department of Forensic Science, Vivekananda Global University, India

Jyoti Pandey
Department of Chemistry, School of Physical and Decision Sciences, B. R. Ambedkar University (A Central University), India

Pritam P. Pandit
Department of Forensic Science, Vivekananda Global University, India

Ankita Patel
Department of Biochemistry and Forensic Science, Gujarat University, India

Amit Kumar Pradhan
Department of Botany, Pragjyotish College, India

Rakesh Rawal
Department of Life Science, Gujarat University, India

Preetom Regon
Department of Botany, Gauhati University, India

Saurov Jyoti Roy
Department of Botany, Gauhati University, India

Anuradha Sandhu
Department of Forensic Science, School of Bioengineering and Biosciences, Lovely Professional University, India

Mahipal Singh Sankhla
Department of Forensic Science, University Center for Research and Development (UCRD), Chandigarh University, Mohali, Punjab, India

Anuj Sharma
Department of Forensic Science, Vivekananda Global University, India

Hanoor Sharma
Clear Labs, California, United States

Manju Shri
School of Applied Sciences and Technology, Gujarat Technological University, Ahmedabad, Gujarat, India

Apoorva Singh
Department of Forensic Science, Vivekananda Global University, India

Gaurav Kumar Singh
Department of Forensic Science, Sharda School of Allied Health Sciences (SSAHS), Sharda University, Greater Noida, Uttar Pradesh, India

Jitender Singh
Institute of Pharmaceutical Sciences (IPS Bhaddal), IET Bhaddal Technical Campus, Ropar, Punjab, India

Pradyumna Kumar Singh
Council of Scientific and Industrial Research–National Botanical Research Institute (CSIR-NBRI), Lucknow, Uttar Pradesh, India; School of Earth and Environmental Sciences, Babasaheb Bhimrao Ambedkar University, Lucknow, Uttar Pradesh, India

Prashant Singh
School of Medico-Legal Studies, National Forensic Sciences University, India

Sudhakar Srivastava
Plant Stress Biology Laboratory, Institute of Environment and Sustainable Development, Banaras Hindu University, India

Bhaben Tanti
Department of Botany, Gauhati University, India

Hunmily Teronpi
Department of Botany, North Gauhati College, India

Rudra Deo Tripathi
Council of Scientific and Industrial Research–National Botanical Research Institute (CSIR-NBRI), Lucknow, Uttar Pradesh, India; Academy of Scientific and Innovative Research (AcSIR), Ghaziabad, Uttar Pradesh, India

Rohit Kumar Verma
Dr. APJ Abdul Kalam Institute of Forensic Science and Criminology, Bundelkhand University, India

Shikha Verma
French Associates Institute for Agriculture and Biotechnology of Drylands, The Jacob Blaustein Institutes for Desert Research, Ben-Gurion University of the Negev, Sede Boqer Campus, Israel

Abbreviations

AAS	atomic absorption spectroscopy
AB	arsenobetaine
ABC	ATP binding cassette
ABCC	C-type ATP-binding cassette transporters family
AC	arsenocholine
Al	aluminum
AM	arbuscular mycorrhiza
AMF	arbuscular mycorrhizal fungi
APX	ascorbate peroxidase
AQP	aquaporins
AR	arsenate reductase
ArsC	cytosolic arsenate reductase
ArsM	adenosylmethionine methyltransferase
As	arsenic
As2S3	arsenic trisulfide
AsA	P-arsanilic acid
AsB	arsenobetaine
AsC	arsenocholine
Asi	inorganic arsenic
ATP	adenosine triphosphate
ATQ1	arsenate tolerance QTL1
ATSDR	Agency for Toxic Substances and Disease Registry
AWD	alternate wetting and drying condition
BCM	biologically controlled mineralization
bHLH	basic helix-loop-helix
BIM	biologically induced mineralization
CaO	calcium oxide
CAT	catalase
Cd	cadmium
CDC25s	cell division cycle protein
CLT	chloroquine resistance transporter
Co	Cobalt
CoA	co-acetyl enzyme

Cr	chromium
Cu	copper
CuO	copper oxide
Cys	cysteine
DHAR	dehydroascorbate reductase
DMA	dimethyl arsenic acid
DMPAO	dimethyl phenyl arsine oxide
DOC	dissolved organic carbon
DPAA	diphenylarsinic acid
DR	Dubinin-Radushkevich
ETAAS	electrothermal atomic absorption spectroscopy
EU	European Union
FAO	Food and Agricultural Organization
Fe	iron
FI HG AAS	flow injection – hydride generation atomic absorption spectroscopy
GE	genetically engineered
g-EC	g-glutamylcysteine
GIT	gastrointestinal tract
GM	genetically modified
GPOX	guaiacol peroxidase
GPX	glutathione peroxidase
GR	glutathione reductase
Grx	glutaredoxin
GS	glutathione synthetase
GSH	glutathione
GST	glutathione S transferase
H_2O_2	hydrogen peroxide
HAC	high arsenic content
HG AAS	hydride generation atomic absorption spectroscopy
Hg	mercury
HMs	heavy metals
HPLC	high-pressure liquid chromatography
HPLC	high-performance liquid chromatography
ICP-MS	inductively coupled plasma mass spectrometry
LA ICP MS	laser ablation Inductively coupled plasma mass spectrometry
LA	laser ablation

LSI	low silicon
Lsi, Lsi1, Lsi2	low silicon protein transporters
MAE	microwave-assisted extraction
MATE	multidrug and toxic compound extrusion
MDHAR	monodehydroascorbate reductase
MDPAO	methyl diphenyl arsine oxide
MetO	methionine sulfoxide
MMA	monomethyl arsenic acid
Mn	manganese
MPAA	methylphenylarsinic acid
MTDI	maximum tolerable daily intake
MTs	metallothionein
N	nitrogen
Ni	nickel
NIP	nodulin 26-like intrinsic protein
NO	nitric oxide
NRAMP	natural resistance-associated macrophage proteins
OAS-TL	o-acetylserine (thiol)-lyase
OVT	ovular vascular trace
P	phosphorus
PAA	phenylarsonic acid
Pb	lead
PC	phytochelatin
PCS	phytochelatin synthase
PHR2	phosphate starvation response 2
PHT1	phosphate transporter 1
Pi	inorganic Phosphate
PIP	plasma membrane intrinsic protein
PO_4^{3-}	phosphate
POD	peroxidase
PvPht	*Pteris vittata* phosphate transporter
RDX	royal demolition explosive
ROL	radial oxygen loss
ROS	reactive oxygen species
RSBC	rice straw biochar
S	sulfur
S2–	sulfide
SAM	S-adenosyl-L-methionine

Si (OH)4	ortho silicic Acid
Si	silica
SiO2	silicon dioxide
SLIM1	sulfur limitation 1
SO4	sulfate
SODs	superoxide dismutases
SPE	solid phase extraction
TBAB	tetra-n-butylammonium bromide
TETRAsv	tetramethyl arsonium cation
TFA	trifluoroacetic acid
TFs	transcription factors
TMA	trimethylarsine
TMAO	trimethylarsine oxide
TNT	trinitrotoluene
U	uranium
USEPA	United States Environmental Protection Agency
UV	ultraviolet
VOC	volatile organic compounds
WHO	World Health Organization
XANES	X-ray absorption near-edge structure
Zn	zinc
ZVI	zero-valent iron

Preface

All around the world, humans are threatened with arsenic by ingesting water, air, and food. It is a global menace as the hassle of arsenic causes anxiety, from young children to the elderly, and the spread of arsenic pollution is unfolding massive areas of nations like Bangladesh. South and Southeast Asian international locations pay excessive prices for arsenic poisoning, and researchers worry that health dangers might be exacerbated by increasing ranges of arsenic pollutants. The possibility of arsenic infection in consuming water, as well as the danger it causes to international health, is much larger than hitherto understood. Arsenic intoxication causes high degrees of skin, lung, and bladder tumors, as well as coronary heart and other lung ailments. This present book emphasizes the overall arsenic problem, especially in rice grown in arsenic-contaminated zones.

The very first chapter of the book describes arsenic in rice in a nutshell. The next chapter deals with the status of arsenic contamination in paddy fields, particularly in the context of the soil water chemistry of arsenic in the field conditions. The third chapter is about rhizospheric physiology, describing changes in the rhizospheric zone occurring under rice cropping in paddy fields. It is very important to know how arsenic is transported inside the plant and how arsenic causes toxicity to rice plants. To deal with these points, the fourth, fifth, and sixth chapters discuss the long-distance translocation of arsenic in rice, which will enable us to understand how arsenic goes inside and what it does to the plant. These chapters also discuss arsenic-induced phytotoxicity in rice plants after entry.

The seventh chapter is all about how economic measures can be taken to reduce the arsenic load inside the rice plant and grains, so it explains several techniques that can be utilized agronomically and are environmentally safe. The eighth chapter present biotechnological and nanotechnological methods that can be utilized to mitigate the arsenic stress in the rice. The ninth chapter is on microbial-mediated techniques that can be applied to regulate arsenic in rice. Last but not least, the final chapter describes all arsenic species present in rice reported to date.

The book provides an excellent report on the issue of arsenic in rice with the latest up-to-date information. The book will be a wonderful resource for students, academicians, and others to gain knowledge on the arsenic-rice problem.

Introduction

Arsenic is a human and animal carcinogen. Arsenicals have been linked to skin, lung, and bladder cancers. Chronic arsenic poisoning also has non-cancerous outcomes such as hyper- and hypopigmentation, keratosis, hypertension, cardiovascular disease, and diabetes. The toxicity of inorganic and organic arsenic compounds is considered to be a significant source of concern for hundreds of millions of people worldwide. Considering the toxic nature of arsenic and its presence in all environmental media and food items, regulatory agencies have issued guidelines for the maximum permissible limits of arsenic in drinking water, rice grains, and rice-based food products. This present book contains three major sections: (i) arsenic in water and soil; (2) arsenic chemistry in rice; and (3) mitigation strategies.

The book provides a comprehensive and latest update on arsenic contamination scenarios in water, soil, and rice, with the latest guidelines and information about arsenic mitigation strategies. Such updated information about arsenic contamination and its spread through food was not available under one umbrella until now.

Further, the book highlights rice with up-to-date mechanistic details of arsenic uptake, transport, and in planta distribution and agronomic and easily feasible measures to tackle the arsenic accumulation in rice grains. The book also discusses human health risks associated with arsenic ingestion and available or possible curative measures that include agronomic, biotechnological, and nanotechnological methods and microbial methods to regulate arsenic levels in rice plants and grains. The book covers the vast horizon of arsenic contamination in the environment.

CHAPTER 1

Arsenic in Rice in a Nutshell: Tell-Tale Toxic Interplay of Chemical Offense and Biochemical Defense

JAYANTA KUMAR BISWAS,[1] RUBINA KHANAM,[2] and SUDHAKAR SRIVASTAVA[3]

[1]*Department of Ecological Studies and International Center for Ecological Engineering, University of Kalyani, India*

[2]*ICAR–National Rice Research Institute, India*

[3]*Plant Stress Biology Laboratory, Institute of Environment and Sustainable Development, Banaras Hindu University, India*

ABSTRACT

Arsenic (As) contamination of the environment and crop plants is an issue of global concern. Public and environmental health is threatened by the excess of As in soil, water, and air and crop produce like rice grains. Rice is one of the major grains consumed by people around the world and rice grains have been demonstrated to accumulate As in high amounts in numerous studies. The problem of As accumulation in rice has gained huge attention among researchers. A lot of research on understanding the uptake and transport mechanisms of As in rice plants and on delineating the mechanism of As toxicity in plants has been conducted. The knowledge about As transporters, regulators of gene expressions, enzymes, and metabolites involved, and anatomical and morphological changes is now available. The vast knowledge is summarized in the present chapter

Arsenic in Rice: A Global Menace. Garima Awasthi, Sudhakar Shrivastava, and Mahipal Singh Sankhla, (Eds.)

to provide up-to-date information about the problem of As in rice, the journey of As from roots to grains, and the mechanism of As toxicity in rice plants.

1.1 ARSENIC IN RICE: A CAUSE FOR CRITICAL CONCERN

Arsenic (As) is one of the most toxic environmental pollutants occurring in organic and inorganic forms. As a highly toxic metalloid and a non-threshold group I carcinogen, As is ranked number one in the Agency for Toxic Substances and Disease Registry's (ATSDR) substance priority list (Martinez et al., 2011). Inorganic arsenic's (Asi) exposure through water and food is a serious global health issue. Chronic exposure to arsenic in drinking water has been related to cancers of the skin, lung, bladder, and prostate, while low-level exposure is linked to various metabolic disorders, obesity, diabetes, liver fibrosis, and cardiovascular disorders.

Arsenic contamination has been documented globally in drinking water and food crops, especially in rice, which affects millions of people worldwide, particularly in India, Bangladesh, China, Taiwan, parts of South America, and Southeast Asia (Raessler, 2018; Shahid et al., 2020). Bangladesh and India possess the largest population at risk due to As exposure. Although the primary reason for arsenic exposure is contaminated groundwater, there are alarming levels of Asi found in the soil. In Bangladesh, about 80% of the As-contaminated land area is used for agriculture, and As uptake by rice is the reality. However, the contamination of soil by As is not limited to groundwater. It can also be attributed to various natural and anthropogenic factors. In China, the soil is polluted by natural biogeochemical processes. Mining and arsenic-rich rocks occurring naturally in Brazil are responsible for As contamination in soil. Whereas, in the United States, As in soil is due to the extensive application of pesticides until the 1980s. Arsenic in contaminated soil is taken up by staple crops, thus increasing the risk of As exposure through food. Barley, legumes, cotton, maize, and rice are commercial crops that have been reported to accumulate As from soil, of which arsenic accumulation in rice has drawn both global attention and apprehension (Golui, 2017; Biswas et al., 2020). Rice plants take up As from soils, and accumulate in the grains, posing human health risks via the soil-water-crop-food pathway (Moulick et al., 2021).

The concentration and speciation of As in soil are regulated by the soil's pH, Eh, organic matter concentration, and nutrient levels, particularly silicon and phosphorus, iron oxide, and clay mineral contents (Shaheen et al., 2017). The As accumulation in rice plants is governed by various biological and geochemical factors which regulate the solubility and bioavailability of As in the plant-soil environment (Zhao et al., 2010; Moulick et al., 2022). It is reported that rice grains accumulate up to 10 folds As than other cereal crops under field conditions which is attributed to the culture (flooded) conditions. Due to high water levels, an anaerobic environment is generated which converts arsenic from bound form to mobile form, leading to higher uptake. As-toxicity enhances the oxidative stress and inhibition of ATP (adenosine triphosphate) formation which leads to a reduction in grain yield. The common symptoms of As toxicity in rice comprise inhibition of seed germination, stunted growth of rice seedlings, brown spots and scorching on leaves, and reduced rate of photosynthesis and grain yield (Moulick et al., 2021; Yadav et al., 2021; Srivastava et al., 2019). Reduction in growth and photosynthetic rate due to As hinders the carbohydrate metabolism in rice and enhances the production of reactive oxygen species (ROS) and hydrogen peroxide (H_2O_2) (Upadhyay et al., 2022; Biswas et al., 2020).

Rice makes up a major part of the diet for about half of the world's population. For many people, it is the only source of carbohydrates, minerals, and some essential vitamins. It is one of the most important food crops in terms of calorie intake and nutrition. Besides being the main staple in many parts of the world, rice has specific health benefits. Rice is recommended for people with celiac disease, in which wheat gluten damages the intestinal lining, and rice acts as a gluten-free substitute in the diet. Although rice is consumed by people all over the world every rice consumer should be concerned about the amount of rice consumed which sets the risk of As exposure (Hojsak et al., 2015; Biswas et al., 2020). Moreover, Asian diets have much more rice in their meals which makes them more susceptible to arsenic-related disorders. The condition in Bangladesh and West Bengal, India is much dire where As exposure affects millions of people. The As content in rice can depend on its variety, place of cultivation, and mode of processing (Moulick et al., 2022). Evidently, rice grown in a hotspot like Bangladesh will have more As content.

To assess the risk of exposure, it is important to know where the rice grains come from. Abedin et al. (2002) found that in the rice seed As accumulation increases in the following order: grain<straw< husk. Thus, the concentration of arsenic in the rice grain could be very less compared to the total arsenic content in the seed. Support for this statement has come from other reports according to which 10–20 times of arsenic is concentrated in the bran layers than the whole grain. It has been reported that brown rice tends to accumulate more arsenic than white because brown rice is not polished and has the bran layers intact. Milling and polishing reduce arsenic content in rice, thus brown rice has the highest arsenic accumulation followed by parboiled brown rice, white rice, and parboiled white rice in decreasing order of arsenic content.

1.2 JOURNEY OF ARSENIC FROM SOIL TO GRAIN

1.2.1 UPTAKE OF ARSENIC SPECIES BY ROOTS

The presence of As species in the rhizospheric soil has a significant impact on As absorption and translocation in rice plants (Martínez-Castillo et al., 2022). Arsenite is more toxic than As(V) among the inorganic forms of As, and their modes of toxicity and transport in plants vary. The reduced environment in the lowland rice production system is congenial for the uptake and translocation of As(III), which exists mostly as uncharged arsenious acid ($H_3AsO_3^0$) at pH 8 (pKa=9.22). The presence of Si and P in soil solution was also found to influence As uptake by rice plants. Arsenate is an analog of phosphates and phosphate transport pathways have been identified for As(V) uptake in rice plants, and new evidence shows that As(III) (silicic acid analog) and undissociated methylated As species dimethylarsinic acid (DMA) and monomethyl arsenic acid (MMA) enter in the root via aquaporin channels, which are also used as silicate transport channels.

1.2.1.1 UPTAKE OF ARSENITE

Too far, many aquaglyceroporins of various classes have been identified for As(III) absorption in rice, most particularly the nodulin 26-like intrinsic protein (NIP) family of aquaporin channels (Pan et al., 2020) (Figure 1.1). Because As(III) is a chemical counterpart of silicic acid, the genes involved in As(III) absorption and transport in rice plants are also involved in silicon (Si) uptake (Awasthi et al., 2017). Till date OsNIP2;1, OsNIP1;1, OsNIP2;2, OsNIP3;1, OsNIP3;2, OsNIP3;3, OsNIP3;3 are NIP class transporters known for silicic acid and As(III) uptake in rice (Table 1.1). In the case of As(III) root absorption, OsNIP2; 1 (OsLsi1), which is primarily present on the distal side of Casparian bands, transports As(III) from soil to root significantly. Recently it was shown that OsNIP3;2 plays an important role in As(III) absorption by rice lateral roots (Chen et al., 2017). The gene OsNIP3;2 is mostly located in lateral roots and the stele area of main roots, and its mutation results in lower As(III) concentrations in roots but not in shoots. Downregulation of messenger RNA production of OsLsi1 in rice by an arbuscular mycorrhizal fungus (AMF) colonization might decrease As(III) absorption and therefore limits As toxicity in rice (Chen et al., 2017). Aquaporins (AQP) of PIP (plasma membrane intrinsic proteins) have also been reported to be involved in As(III) absorption in rice (Kumarathilaka et al., 2019). According to studies (Hussain et al., 2021), the most common PIPs for As(III) absorption in rice are OsPIP2;4, OsPIP2;6, and OsPIP2;7. Other transporters, such as NRAMP1 (Natural Resistance-Associated Macrophage Protein 1), may be involved in As(III) absorption and transport. Because Si and As(III) compete for root absorption, adding Si to the soil has been proven to be beneficial to reduce As levels in rice grains (Khanam et al., 2022).

TABLE 1.1 List of Transporters Involved in the Journey of Arsenic from Soil to Rice Grain

Transporters	Family	Species	Remarks
OsPT1 OsPT2 OsPT8 OsPT4	Pht1 family	As(V)	Soil to root and inside the plant.
OsNIP2;1 (Lsi1) OsNIP1;1 OsNIP3;1 OsNIP3;2 OsNIP3;3	NIPs	As(III), DMA, MMA	Soil to root.
OsNRAMP1	NRAMP	DMA, MMA, As(V)	Soil to root.
OsNIP2;2 (Lsi2)	NIPs	As(III)	Root to shoot.

TABLE 1.1 *(Continued)*

Transporters	Family	Species	Remarks
OsPIP1;2	PIPs	As(III)	Soil to root and root to shoot.
OsPIP1;3			
OsPIP2;4			
OsPIP2;6			
OsPIP2;7			
OsPTR7	Putative peptide	DMA	Xylem and phloem loading.
OsHAC1;1	Other NIPs	As(V)	Reduce As(V) to As(III) in root.
OsHAC1;2			
OsHAC4			
OsNPF8;1			
OsABCC1	ABC	As-PCs	As detoxifying (transport As-PC to vacuoles).
OsABCC1			–
OsNIP2;1	NIP	As(III)	Efflux to the environment.
ArsM	Microbial genome	MMA, DMA, TMA	Methylation of inorganic As species.

1.2.1.2 UPTAKE OF ARSENATE

As discussed, under flooded paddy soil As(III) is the dominant species. However, when rice is grown under non-flooded condition (aerobic rice), it predominantly uptake As(V). Moreover, the oxidized rhizosphere of rice also facilitates the conversion of As(III) to As(V). Under such circumstances, the most common form of As in soil is As(V), which has a similar structure to PO_4^{3-}. It has been reported that efficient As(V) uptake in rice is most likely accomplished by sharing a number of phosphate transporters such as OsPHT1;1 and OsPHT1;8. Ye et al. (2017) discovered OsPT4, belonging to the Pht1 family can also efficiently uptake As(V). Wang et al. (2016) compared the As(V) tolerance of an indica variety Kasalath to that of the japonica variety Nipponbare and found that Kasalath was more tolerant than Nipponbare. This might be linked to 2- to 3-fold overexpression of OsPT2 and OsPT8 in Nipponbare. The downregulation of OsPHT both in Kasalath and Nipponbare demonstrated of 33–57% reduction in As(V) absorption and showed enhanced tolerance to As(V). As a result, Pht1 family transporters were identified as the key transporters for As(V) absorption in rice.

1.2.1.3 UPTAKE OF METHYLATED ARSENIC SPECIES

Researchers confirmed that microbiological conversion of inorganic species in submerged paddy soil leads to generate organic As species, specifically monomethyl arsenic acid (MMA; $CH_3AsO(OH)_2$) and dimethylarsinic acid (DMA; $(CH_3)_2AsOOH$). Remarkably, Lsi1 has been found to be involved in MMA and DMA absorption. Regardless, Lsi2 was not found to be DMA permeable. The absorption efficiency of organic (methylated) As species is lesser than inorganic As species, and it reduces further as the number of methyl groups increases (Abedi & Mojiri, 2020). Furthermore, in rice roots, MMA(V) is predominantly transformed to MMA(III), restricting its translocation to the shoot.

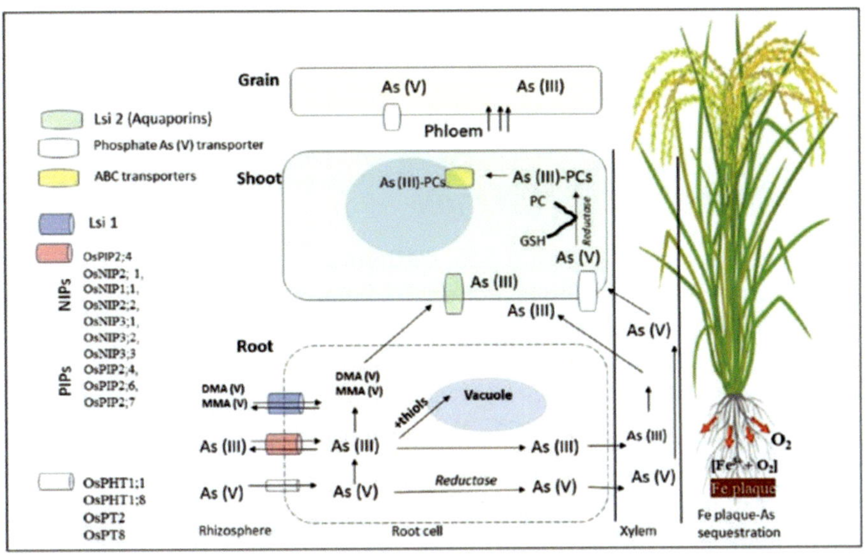

FIGURE 1.1 Summarized diagram of the journey of arsenic from soil to grain.

1.2.2 TRANSFORMATION OF ARSENIC SPECIES IN PLANT

1.2.2.1 ARSENATE REDUCTION

The transformation of As(V) to As(III) in plants is regulated by arsenate reductase (AR) enzymes. According to Awasthi et al. (2017), As(V) reduction occurs fast in all root apex tissues, and As(V) is the minor species found exclusively in the rhizodermis. Several AR genes have been identified in plants, however, their role in As(V) reduction is uncertain. In rice, recently HAC1 viz., OsHAC1;1 and OsHAC1;2 (orthologous genes) has been identified to operate as As(V) reductases and are expressed predominantly in roots (Fischer et al., 2020). However, their distribution is diverse. For example, OsHAC1;1 is predominantly expressed in the epidermis, root hairs, and pericycle while OsHAC1;2 is expressed in the epidermis, outer cortex layers, and endodermis. Further, Xu et al. (2017) discovered that OsHAC1;1 is also expressed in the stems and nodes. HAC4 was recently found by Xu et al. (2017) as an As(V) reductase from rice which is predominantly expressed in root elongation and maturation zones of the epidermis and exodermis. The downregulation of OsHAC1;1, OsHAC1;2, and OsHAC4 led to a decrease in As(V) reduction in roots and decreased As(III) efflux, and increased As(V) accumulation in shoots. Overexpression of these genes generates opposing consequences. In comparison to other cereal crops, the increased efficiency of As(V) reduction in rice roots, followed by efficient loading of As(III) to xylem sap, contributes significantly to high grain As(III).

1.2.2.2 ARSENITE COMPLEXATION

Arsenite has a strong affinity to sulfhydryl (–SH) groups of peptide thiol like glutathione (GSH) and phytochelatins (PCs), hence As(III) is detoxified in plants by complexing with GSH or PCs (Martínez-Castillo et al., 2022). Glutathione is synthesized in a two-step process catalyzed by the rate-limiting g-glutamylcysteine (g-EC) synthetase to synthesize g-EC, combing Gly and g-EC. Phytochelatins are synthesized through the transpeptidation of gamma-glutamyl cysteinyl dipeptides from GSH by the catalytic activity of phytochelatin synthase (PCS) (Martínez-Castillo et al., 2022). The findings revealed that the translocation mobility of organic As species is better than inorganic As species. This has been linked to the complexation and storage of As(III) mediated by PCs. Although uncomplexed As(III) is highly mobile and dominant in xylem and phloem transport, As(III)-thiol complex formed in rice roots is assumed to limit root-to-shoot As(III) translocation.

1.2.2.3 ARSENITE EFFLUX

Plants may quickly convert As(V) to As(III) in their roots, which can subsequently be effluxed into the environment (Mawia et al., 2021).

Improving As(III) efflux through plant roots might be a way to reduce As buildup in plants. The major membrane transporters involved in As(III) efflux in plant roots have yet to be identified. Lsi1 (OsNIP2;1), a rice aquaporin that facilitates As(III) absorption and imparts As accumulation, also mediates As(III) efflux, accounting for 15–20% of total As efflux.

1.2.2.4 ARSENIC METHYLATION

Though As methylation is common in microbes and humans, it is uncertain if As methylation occurs in plants *in vivo*. Keren et al. (2021) showed a gene ArsM from the bacterium *Rhodopseudomonas* sp. by studying microbial genomes. They discovered that RpArsM is involved in the synthesis of a variety of methylated intermediates (DMA^V and TMA^O) from As(III). Since rice is incapable to methylate As *in vivo*, methylated As species in rice grains are most likely from soils. ArsM gene in rice may accelerate As methylation, therefore, giving a technique to minimize the accumulation of As in rice grains. Meng et al. (2011) introduced the RpArsM gene into the rice and observed that transgenic rice generates methylated As species and gives out 10-fold higher organic As species over the control. The results also showed that inorganic As accumulation decreases in the first progeny of transgenic rice grains.

1.2.3 SEQUESTRATION OF ARSENIC IN PLANT

According to recent findings, As is largely sequestered in the vacuoles of the endodermis, pericycle, and xylem parenchyma cells. As also sequestrated on the rice root by binding via Fe-plaque.

1.2.3.1 IRON PLAQUE FORMATION

Rice being a semi-aquatic plant with a lot of aerenchyma in its roots transfers a part of the O_2 to the rhizosphere to adapt to the anaerobic conditions in submerged soil. This release of O_2 in the rhizosphere is termed radial oxygen O_2 loss (ROL). The ROL varies with cultivars, the waterlogging condition, and the availability of O_2 in the submerged soils. The release of O_2 facilitates the conversion of ferrous iron (Fe_2O) to ferric iron (Fe_3O)

and leads to the precipitation of iron (hydro) oxides (FeOOH) on the root surface. This thin brown layer of iron (hydro) oxides on the root surface of aquatic plants are characterized as iron plaque (Fe-plaque). Interestingly, Fe-plaque has some functional groups (viz., OH, COOH, etc.) which react with As and act as scavengers of As on rice roots, restricting its entry inside the plants (Khanam et al., 2022). The As content on the Fe-plaque is much higher than the As inside the roots and other plant parts. Further, chemical analysis and speciation study of As adsorbed on Fe-plaque has revelated that As(V) is primarily adsorbed on Fe-plaque (Chen et al., 2017). However, the significance of Fe-plaque in As distribution in rice plants is controversial. It is reported to restrict As entry inside the plant in certain investigations (Mawia et al., 2021). Contrastingly, Fe-plaque is also blamed for supplying As to the root transporters more easily as the Fe-plaque layer is closely lying on the rice roots (Liu et al., 2006). The function of Fe-plaque in affecting As sequestration may also change with its thickness and structure. The thickness and structure of Fe-plaque actually fluctuate from root tips to old roots and from primary to lateral roots. Moreover, a cultivar with higher ROL has the capacity to increase Fe-plaque development on rice roots and limit As entry inside the plant. Additionally, increased ROL may also convert As(III) to As(V), which is more firmly adsorbed to Fe-plaque.

1.2.3.2 VACUOLAR SEQUESTRATION

Once As(III/V) get inside the rice plants, it predominantly converts to its reduced form, i.e., As(III) that may be transported to vacuoles following complexation with PCs. After being chelated by PCs, the PC-As(III) complexes are transported to vacuoles through the ABC transporter family, notably ABCC1 and ABCC2 in rice (Pan et al., 2020). The ABC transporters, located in the tonoplast of phloem cells in nodes, facilitate PC-As(III) complex transport to vacuoles and limits the translocation of As into grains. In the translocation of As, nodes operate as a key hub as they govern the elemental distribution. It has been found that, with significantly greater As concentrations than internodes and other plant parts, rice nodes are the most critical area for As sequestration, functioning as a sieve preventing As transmission to the upper plant parts. Thounaojam et al. (2021) also revealed that the downregulation of the OsABCC1 transporter increased As accumulation in grains but decreased As in nodes. Thus, it is clearly demonstrated that vacuoles present in the nodes, internodes, and roots govern the As storage and restricts its translocation to the rice grain.

1.2.4 TRANSLOCATION (TRANSPORT) OF ARSENIC AND DELIVERY TO RICE GRAIN

On entry to roots, As moves into the shoots, and eventually into the grains. It's been discovered that a silicic acid transporter (OsLsi2) is involved in the transfer of As(III) from the root to the shoot, as well as xylem loading. OsLsi2, which is found on the proximal side of Casparian bands, transfers As(III) from root cells to the apoplast and eventually to the xylem. This Lsi 2-led As transport was predicted to decrease with the addition of Si (Mawia et al., 2021). For example, when Si was applied under flooded conditions, the content of As(III) in polished rice was lowered by 33% as a result of a 50% reduction in the vascular transport of As (Fleck et al., 2013). Arsenate shares the transporters for their translocation inside the rice with phosphate (P). Under aerobic conditions, the phosphate transporter (OsPht1;8) was expressed constitutively, which enhanced the xylem loading of As(V). The majority of the As(V), As(III), and MMA accumulated in the root, but DMA was easily translocated to the shoot (Awasthi et al., 2017). Amongst the organic and inorganic As species, DMA(V) is highly mobile in rice plants, quickly transferring from root to shoot. It has been identified that the OsPTR7 gene is particularly involved in the transport of DMA in the xylem and phloem. According to Muehe et al. (2020), methylated As is taken up more slowly in rice than inorganic species, yet it is readily translocated to grains. As previously stated, thiol complexation reduces As(III) mobility, whereas organic As species have no such constraint for movement. Arsenic unloading into rice grain is of particular significance since it is the most common route of As exposure. As compared to the xylem route, phloem-derived As has been shown to contribute much more to grain As accumulation. There is still a lot to reveal about the transporters that help the translocation of As from the xylem to the phloem and into the seeds. It has been observed that, like other nutrients, As is delivered to the endosperm through chalaza to nucellar projection via the ovular vascular trace. A recent study on As unloading into rice grain found that DMA^{V} is more mobile than As(III) in both phloem and xylem and is translocated to the rice grain with 10 times better efficiency than inorganic species (Carey et al., 2020). Furthermore, Carey et al. (2020) discovered that inorganic As is weakly remobilized from flag leaves to grain via phloem transport, but DMA^V and MMA^V are re-translocated easily. Inorganic and organic As have different mobility patterns, which may be seen in their distribution patterns. According to Punshon et al. (2018), inorganic As is mostly located in the OVT area, whereas DMA is found in the endosperm of rice grains. Furthermore, Sun et al. (2008) found that inorganic As is present in the bran layer of brown rice, but the endosperm contains mainly DMA.

1.3 TOXICITY OF ARSENIC IN RICE PLANT

The effects of As on plants are multifaceted and can bring about changes at morphological, physiological, and biochemical levels (Ahsan et al., 2008; Shri et al., 2009). Germination is a key process in the establishment of healthy plants and the presence of high As around seeds can inhibit seed germination (Shri et al., 2009) and reduce the growth of the seedlings (Yadav et al., 2021). The negative effects at the early growth phase of the plant negatively affect its responsiveness to stress (Srivastava et al., 2013a). One of the key processes sustaining plant life is photosynthesis capturing light energy into chemical energy and to generate biomass. A number of studies in different plants demonstrate that As affects photosynthesis and implicate this to be a major cause of growth retardation (Srivastava et al., 2013b; Mishra et al., 2014). Carbon metabolism is also affected by As accumulation in plants. A recent report by Mishra et al. (2014) demonstrated a decrease in photosynthetic pigments with an increase in As. However, they noticed that core photosynthesis components were affected only when the As concentrations were high. In several transcriptomic and proteomic studies, photosynthesis-related genes and proteins have been found to be mainly affected in response to As stress, and important ones include triosephosphate isomerase, the ribulose-1,5-bisphosphate carboxylase/oxygenase large subunit, oxygen-evolving complex protein 1, etc. (Liu et al., 2013; Chauhan et al., 2020; Srivastava et al., 2021; Awasthi et al., 2021). Srivastava et al. (2021) suggested the role of photosynthesis-related genes to be the primary player behind thiourea-mediated As stress amelioration in rice.

Water and nutrient uptake and homeostasis are also influenced by As toxicity in plants (Srivastava et al., 2013c). Water uptake is mediated by AQP, channel proteins allowing passage of water molecules. However, a subfamily of AQP, namely aquaglyceroporins, is involved in the uptake of small neutral molecules including As in the form of As(III). It has been observed that plants tend to suppress the expression of aquaglyceroporins presumably to reduce As(III) uptake and in the process eventually affect water uptake also (Srivastava et al., 2013c). In the form of As(V), As competes with the uptake and transport of phosphate and can therefore affect phosphate homeostasis and the processes dependent on phosphates like ATP synthesis and phosphorylation reactions (Awathi et al., 2017). The analysis of other elements like Zn, Cu, Fe, Mn, K, etc., has demonstrated the effects of As on their uptake and accumulation also. These effects have been attributed to the changes in the expression of transporters, to the negative effects on root growth and architecture, to the overall reduction of plant growth and altered requirement of nutrients, and changes in chemical species in soil due to reactions of various elements with As (Srivastava et al., 2022). In essence, the disturbed water and nutrient homeostasis lead to changes in the growth of plants.

Arsenic-induced changes in the composition of essential and non-essential amino acids have also been reported (Dwivedi et al., 2012; Pathare et al., 2013). Amino acids are the building blocks of proteins and also certain peptides, which function as the backbone of plant growth. Some amino acids are also crucial from the point of tolerance to stresses, such as histidine for Ni, proline for metal(loid)s, salinity, and drought, and cysteine for metal(loid)s. Among important peptides, glutathione (GSH) and GSH-derived peptides, phytochelatins (PCs) are central players in As detoxification in plants. GSH and PCs are made up of glutamate, cysteine, and glycine with cysteine playing the role of As complexation. The As-induced synthesis of GSH and PCs also exerts pressure on the synthesis of glutamate, cysteine, and glycine (Srivastava & D'Souza, 2009). The effects of As on sulfur assimilation and cysteine metabolism are demonstrated in a number of plants (Mishra et al., 2009; Srivastava et al., 2016). In addition, changes in glutamate and glycine metabolism are also reported (Srivastava & D'Souza, 2009). Apart from synthesis, protein levels are also changed owing to protein oxidation via reactive oxygen species (ROS), whose production is induced profoundly upon As stress. The ROS production is suggested to increase upon As entry owing to the consumption of GSH in As complexation; GSH is a crucial metabolite involved in ROS scavenging directly and via the ascorbate-glutathione cycle (Mylona et al., 1998). Further, the conversion of As from one to another form viz., As(V) to As(III) by reductases consumes reductants and affects ROS metabolism. The decreased activities of antioxidant enzymes while the increase in that of prooxidant enzymes like NADPH oxidase and ascorbate oxidase (Srivastava et al., 2011) can also cause an increase in ROS levels. The disturbed energetic status visualized in terms of ATP/ADP ratio and changes in metabolites of redox viz., NADH/NAD, NADPH/NADP, GSH/GSSG and ascorbate to dehydroascorbate also acts as the major factor of As stress to plants (Srivastava et al., 2011, 2013a). The ROS induced upon As stress leads to oxidative conditions in the cellular environment, causing oxidation of lipids in membranes and damage to proteins and DNA also. Oxidative stress and As toxicity have been suggested to be positively correlated with each other (Srivastava et al., 2011). The lipid peroxidation causes membrane damage and makes them leaky as indicated by increased electrical conductivity. Thus, membrane fluidity and functionality are altered upon As-induced oxidative stress and these in turn cause changes in cell growth. The effects of As on DNA have been visualized via cytological analysis of chromosomal abnormalities in plants (Gupta et al., 2020, 2021). The produced ROS, during As-stress, can cause oxidative base damage in DNA molecules via DNA-protein cross-links, breakage of the chromosome, chromosomal aberrations, aneuploidy, the formation of apyrimidinic/apurinic sites, sister chromatid exchange, and micronuclei formation (Abbas et al., 2018). The amino acids like Arg, His, Lys, and Trp, form a free carbonyl group through the oxidation of amino acid side chains to form aminoketobutyrate or amino adipic semialdehyde (Abbas et al., 2018).

The high concentration of As in the soil decreases the potential of biological nitrogen fixation of leguminous crops and reduces root nodules by 50%, and these effects have been linked to changes in associated gene expressions (Lafuente et al., 2010). Arsenic also disrupts nitrogen assimilation which is important for non-legumes to meet with their N-requirement (Shahid et al., 2019). As(V)-treated rice roots had shown the downregulation in the transcripts for a nitrate transporter and an ammonium transporter (Norton et al., 2008). As(V)-treated roots showed lesser production of GS protein; this can be attributed to the fact that As-toxicity reduces the activities of both nitrate and nitrite reductase which help in supplying NH^{4+} to GS for N assimilation (Abbas et al., 2018; Shahid et al., 2019). Arsenic exposure can alter the expression of the sulfate transporter gene in plants (Norton et al., 2008; Srivastava et al., 2009). Arsenic-induced changes in the expression of miR395 have also been reported (Srivastava et al., 2013d). miR395 is a regulator of sulfur metabolism and can alter the expression of SULTR2;1, and ATP sulfurylases 1, 2, 3 (APS 1, 2, 3). Plants take sulfur in the form of sulfate, which converts into sulfide, further, this gets combined with O-acetyl serine to form Cysteine. This reaction is catalyzed by an enzyme O-acetyl serine (thiol)-lyase (OAS-TL), or cysteine synthase. Arsenic exposure can down-regulate OAS-TL activity in As-sensitive plants as it adversely impacts the biosynthesis of cysteine, GSH, and PCs in the roots (Wang et al., 2020). The knockout lines OsOASTL-A1, in rice, have shown the lower level of Cys, GSH, and PCs in the root and thus making the plant extremely sensitive toward As(V) stress (Wang et al., 2020). Sulfur limitation 1 (SLIM1), a key transcription regulator in the sulfur assimilation pathway, is also get triggered with As exposure. In a study conducted by Jobe et al. (2021), slim1 mutants of *Arabidopsis thaliana*, were found to be sensitive to As, these mutant lines were totally impaired in accumulating thiols and sulfur.

At the regulatory level, As toxicity is presumed to operate through changes in the expression of genes and proteins, which have been revealed now in a number of plants including rice (Awasthi et al., 2021; Chauhan et al., 2020; Abercombie et al., 2008, Norton et al., 2008; Chakraborti et al., 2009; Yu et al., 2012; Sharma et al., 2015; Requejo & Tena, 2005, 2006; Ahsan et al., 2008, 2010; Liu et al., 2013). The pathways affected include transcription factors and phytohormones. The changes in phytohormones like auxins, jasmonic acids, abscisic acid, salicylic acid, etc., have been visualized via changes in gene expressions and also through quantitative analysis of their levels (Srivastava et al., 2021). The mutual interaction of hormones and As is demonstrated to bring changes in plant growth (Srivastava et al., 2013d). A recent study by Yadav et al. (2021) threw light on the metabolomics changes in rice upon As stress. This was the first just comprehensive report on As-mediated metabolomic changes. The work reported variable changes in metabolites upon As stress in a tolerant Pooja and moderately sensitive CO-50 variety.

The prolonged toxicity of As also causes changes at ultrastructure levels. The study by Upadhyay et al. (2022) demonstrated changes in xylem vessels and cellular arrangement in As-exposed rice plants. Similarly, Yadav & Srivastava (2021) visualized change in roots upon As stress and found major changes in root structure and cellular integrity, with collapsed aerenchyma. In other plants also, As induced ultrastructural changes in chloroplasts (Li et al., 2006), mesophyll cells of the cortex root and in cellular differentiation of vessel elements (Singh et al., 2007), in intracellular spaces, vesicles, cortical, and epidermal structures (Scheneider et al., 2013) have been noticed. It was observed that with increasing concentration of As, the amount of starch grains and amyloplasts decreased and the amount of lipids increased. There was a reduction in. A large number of genetic modifications have also come into with the advancement in As-related studies.

Zheng et al. (2013) reported DMA-induced abnormal florets before flowering and a sharp decline in the seed setting rate after flowering in rice. It was found that DMA tended to accumulate in caryopsis and induced higher toxicity to the reproductive tissues resulting in markedly reduced grain yield. Srivastava et al. (2021) recently noticed a unique change in root architecture upon As stress imposition. They demonstrated that crown root number was increased upon As stress in rice. In similar lines, Srivastava et al. (2019) found that root length, and number of lateral roots and root hairs numbers in As stress and in As-nitrogen interaction in rice seedlings.

Arsenic toxicity is also dependent on the species of As. The earlier instrumental facilities were able to quantify only a few species of As but today several As chemical species have been identified. Various As species include inorganic As(III), As(V), and organic DMA, MMA, TMA, thiolated arsenicals (Cullen et al., 2016), arsenosugars, arsenolipids, etc. (Xue et al., 2019). The variation in As toxicity of different species is due to their different reactivity. For example, As(III) reacts more strongly with –SH-containing metabolites and proteins and mainly affects protein functions. AsIII also binds with thiol groups of proteins and cofactors and either reduces or inhibits their activity. As(V) acts mainly through competitive inhibition of phosphate-requiring reactions (Rosen et al., 2011; Tawfik & Viola, 2011). The unstable nature of the AsV-esters hence creates futile reaction cycles that uncouple respiratory and photosynthetic ETC (Finnegan & Chen, 2012).

1.4 ARSENIC STRESS TOLERANCE

The strategies adopted by plants to combat the stress induced by As have been depicted in terms of primary and secondary detoxification strategies (Srivastava et al., 2007). The primary target of plants is to reduce the level of free As ions or to reduce the reactive As species in the cellular environment and this is achieved by complexation with cell wall components or by the specific peptides like phytochelatins, which after binding As sequester to vacuoles. In recent years, it has been found that the higher the amount of pectin, hemicellulose, and pectin demethylation, the greater the As deposition on the roots resulting in lower root-shoot translocation in plants (Ye et al., 2017; Huang et al., 2021). The role of PCs and vacuolar As sequestration is well known to reduce As concentrations in the cellular environment and to reduce As root-to-shoot and shoot-to-grain translocation in rice. In rice, the nodes have been suggested to play a major role in restricting As upward movement and this is achieved by As complexation with PCs (Chen et al., 2015). Secondary detoxification means the strategies targeted to reduce stress caused by As at any point of time during the exposure. In this connection, the role of antioxidants involved in the regulation of ROS production becomes crucial. If the ROS are not regulated effectively, the stress worsens rapidly. There are a number of molecular and enzymatic antioxidants that play a role in different cellular compartments and tissues to counteract oxidative stress.

Among ROS, superoxide radicals are generated when electrons are misdirected toward oxygen molecules. With the subsequent electron transfer, hydrogen peroxide and hydroxyl radicals are produced. Superoxide dismutases (SODs) are proposed to be at the forefront of antioxidant defense due to their action on superoxide radicals. SODs dismutate superoxide radicals to hydrogen peroxide and water. The metal cofactors perform the function of electron donor during superoxide dismutation and SODs have associated themselves with different metals during the course of evolution. On the basis of metal cofactors, SODs are classified into Fe-SOD, Mn-SOD, and Cu-Zn-SOD, which are located in different cellular organelles and a particular organelle, either a single type or all types of SODs can exit (Alscher et al., 2002).

H_2O_2 produced is neutralized by a number of peroxidases and catalases (Srivastava et al., 2007). Catalase converts H_2O_2 to water molecules and does not involve any other reductant in the process. Peroxidases perform the same function but need the presence of other metabolites in the process, such as ascorbate for ascorbate peroxidase (APX) and glutathione for glutathione peroxidase (GPX). Among various peroxidases, APX constitutes the most important enzyme that operates through ascorbate-glutathione cycles for the regulation of H_2O_2 levels. The ascorbate-glutathione cycle comprises several enzymes namely APX, dehydroascorbate peroxidase

(DHAR), monodehydroascorbate peroxidase (MDHAR), glutathione reductase (GR), and metabolites viz., ascorbate, glutathione, and NADPH. The ultimate reductant in the cycle is NADPH. Ascorbate and glutathione playing a role in the cycle can also quench different ROS by direct reaction as well and are therefore important molecular antioxidants of the cells. Other non-enzymatic molecular antioxidants comprise carotenoids, flavonoids, phenolics, proline, etc. These antioxidants help in the regulation of several peroxides other than H_2O_2 also and help maintain cellular integrity and membrane structure intact. A number of studies have demonstrated the accumulation of antioxidants during As exposure (Upadhyay et al., 2022; Yadav et al., 2021; Srivastava & D'Souza, 2010). A role of glutaredoxins (GRXs) is also demonstrated in As stress and these may help in ROS regulation via GSH and also in As(V) reduction to As(III) for the purpose of its complexation (Sundaram et al., 2009).

1.5 CONCLUSION

- Arsenic cause phytotoxic effects enhancing oxidative stress, inhibition of ATP synthesis, and reduction of seed germination, photosynthesis, plant growth, and grain yield. Arsenic could appear as a double-edged sword as it can depress crucial micronutrients such as selenium in rice grains.
- Rice plant is capable of reducing As-induced stress to a certain extent adopting some strategies and mechanisms including: (i) eliminating As from cells involving specific transporters or compartmentalization; (ii) inducing the members of the plant defense repertoire (both enzymatic and non-enzymatic) to be hyperactive for facing and mastering the challenge of the free radicals/reactive oxygen species which otherwise cause cellular injury; and (iii) detoxifying As-induced toxicity by the production and/or accumulation of compatible solutes or osmolites such as glycinebetaine, proline, and mannitol.
- Accumulation of As in rice and consequent transmission into the food web can pose serious health risks to the people dependent on rice and rice-based products. The threat is not limited to the people living in the As-hot spot areas, but the threat can spread to the areas where the contaminated rice is transported or imported for consumption.

- Different strategies adopted to curtail the arsenic menace revolve around mainly two categories of approaches, namely agronomic and biotechnological. Agronomic approaches entail amending fertilizers through supplementation of nutrients such as Fe, S, Si, and Se (that can reduce the effect of As), adopting water management practices, and the use of biological agents (e.g., rhizospheric bacteria *Bacillus*, *Rhodococcus*, *Halobacterium,* etc.). Biotechnological approaches comprise the production of transgenic rice varieties by altering the expression of transporters and genes involved in As uptake, translocation, and sequestration throughout the plant, enhancing the synthesis of GSH and PCs and, thus, raising methylation and volatilization of As, and the use of plant breeding techniques. The best strategy is the multipronged approach blending the best possible means: (i) targeting the reduction of As load in rice plus improving levels of certain nutrients including phosphate and silicic acid (transporters of which are involved in As uptake); (ii) modification in the expression level of transporters and genes which can reduce As in different plant parts (roots, shoots, and edible parts), but it should be monitored properly so that these modifications do not alter the expression level of other genes responsible for uptake and transport of other essential minerals or nutrients; and (iii) limiting As to soil and reducing its uptake by roots as an effective approach if the composition of the soil, its nutrients status and microbial profile are properly maintained. Above all, the quality of rice grains needs to be conserved under any management or manipulative intervention.

KEYWORDS

- **arsenic**
- **carbohydrates**
- **cardiovascular disorders**
- **fibrosis**
- **hydrogen peroxide**
- **metabolic disorders**
- **rice seedling**

REFERENCES

Abbas, G., Murtaza, B., Bibi, I., Shahid, M., Niazi, N. K., Khan, M. I., Amjad, M., Hussain, M., & Natasha, (2018). Arsenic uptake, toxicity, detoxification, and speciation in plants: Physiological, biochemical, and molecular aspects. *Int. J. Environ. Res. Public Health, 15*, 59.

Abedi, T., & Mojiri, A., (2020). Arsenic uptake and accumulation mechanisms in rice species. *Plants, 9,* 129.

Abedin, M. J., Cresser, M. S., Meharg, A. A., Feldmann, J., & Cotter-Howells, J., (2002). Arsenic accumulation and metabolism in rice (*Oryza sativa* L.). *Environ. Sci. Technol., 36*, 962–968.

Abercombie, J. M., Halfhill, M. D., Rajan, P., Rao, M. R., Saxton, A. M., Yuan, J. S., & Stewart, C. N. Jr., (2008). Transcriptional responses of *Arabidopsis thaliana* plants to As(V) stress. *BMC Plant Biol., 8*, 87.

Ahsan, N., Lee, D. G., Alam, I., Kim, P. J., Lee, J. J., Ahn, Y. O., Kwak, S. S., et al., (2008). Comparative proteomic study of arsenic-induced differentially expressed proteins in rice roots reveals glutathione plays a central role during As stress. *Proteomics, 8*, 3561–3576.

Ahsan, N., Lee, D. G., Kim, K. H., Alam, I., Lee, S. H., Lee, K. W., Lee, H., & Lee, B. H., (2010). Analysis of arsenic stress-induced differentially expressed proteins in rice leaves by two-dimensional gel electrophoresis coupled with mass spectrometry. *Chemosphere, 78*, 224–231.

Alscher, R. G., Erturk, N., & Heath, L. S., (2002). Role of superoxide dismutases (SODs) in controlling oxidative stress in plants. *J. Exp. Bot., 53*, 1331–1341.

Awasthi, S., Chauhan, R., Indoliya, Y., Chauhan, A. S., Mishra, S., Agrawal, L., Dwivedi, S., et al., (2021). Microbial consortium mediated growth promotion and arsenic reduction in rice: An integrated transcriptome and proteome profiling. *Ecotox. Environ. Saf., 228*, 113004.

Awasthi, S., Chauhan, R., Srivastava, S., & Tripathi, R. D., (2017). The journey of arsenic from soil to grain in rice. *Front. Plant Sci., 8*, 1007.

Biswas, J. K., Warke, M., Datta, R., & Sarkar, D., (2020). Is arsenic in rice a major human health concern? *Curr. Poll. Rep., 6*, 37–42.

Carey, M., Meharg, C., Williams, P., Marwa, E., Jiujin, X., Farias, J. G., De Silva, P. M., et al., (2020). Global sourcing of low-inorganic arsenic rice grain. *Exposure Health, 12*, 711–719.

Chakrabarty, D., Trivedi, P. K., Misra, P., Tiwari, M., Shri, M., Shukla, D., Kumar, S., et al., (2009). Comparative transcriptomic analysis of arsenate and arsenite stresses in rice seedlings. *Chemosphere, 74*, 688–702.

Chauhan, R., Awasthi, S., Indoliya, Y., Chauhan, A. S., Mishra, S., Agrawal, L., et al., (2020). Transcriptome and proteome analyses reveal selenium mediated amelioration of arsenic toxicity in rice (*Oryza sativa* L.). *J. Hazard. Mater., 390*, 122122.

Chen, Y., Han, Y. H., Cao, Y., Zhu, Y. G., Rathinasabapathi, B., & Ma, L. Q., (2017). Arsenic transport in rice and biological solutions to reduce arsenic risk from rice. *Front. Plant Sci., 8*, 268.

Chen, Y., Moore, K. L., Miller, A. J., McGrath, S. P., Ma, J. F., & Zhao, F. J., (2015). The role of nodes in arsenic storage and distribution in rice. *J. Exp. Bot., 66*, 3717–3724.

Cullen, W. R., Liu, Q., Lu, X., McKnight-Whitford, A., Peng, H., Popowich, A., Yan, X., et al., (2016). Methylated and thiolated arsenic species for environmental and health research — A review on synthesis and characterization. *J. Environ. Sci., 49*, 7–27.

Dwivedi, S., Mishra, A., Tripathi, P., Dave, R., Kumar, A., Srivastava, S., Chakrabarty, D., et al., (2012). Arsenic affects essential and non-essential amino acids differentially in rice grains: Inadequacy of amino acids in rice based diet. *Environment International, 46*, 16–22.

Finnegan, P. M., & Chen, W., (2012). Arsenic toxicity: The effects on plant metabolism. *Front. Physiol., 3*, 182.

Fischer, S., Sánchez-Bermejo, E., Xu, X., Flis, P., Ramakrishna, P., Guerinot, M. L., Zhao, F. J., & Salt, D. E., (2021). Targeted expression of the arsenate reductase HAC1 identifies cell type specificity of arsenic metabolism and transport in plant roots. *J. Exp. Bot., 72*, 415–425.

Fleck, A. T., Mattusch, J., & Schenk, M. K., (2013). Silicon decreases the arsenic level in rice grain by limiting arsenite transport. *J. Plant Nutr. Soil Sci., 176*, 785–794.

Golui, D., Guha, M. D. N., Sanyal, S. K., Datta, S. P., Ray, P., Patra, P. K., Sarkar, S., & Bhattacharya, K., (2017). Safe limit of arsenic in soil in relation to dietary exposure of arsenicosis patients from Malda district, West Bengal – A case study. *Ecotox. Environ. Saf., 144*, 227–235.

Gupta, K., Srivastava, S., Saxena, G., & Kumar, A., (2022). Application of *Pteris vittata* L. for phytoremediation of arsenic and biomonitoring of the process through cyto-genetic biomarkers of *Trigonella foenum-graecum* L. *Physiol. Mol. Biol. Plants, 28*, 91–106.

Gupta, K., Srivastava, S., Saxena, G., & Kumar, A., (2022). Evaluation of phytoremediation potential of *Pteris vittata* L. on arsenic contaminated soil using *Allium cepa* bioassay. *Bull. Environ. Contam. Toxicol., 108*, 423–429.

Hojsak, I., Braegger, C., Bronsky, J., Campoy, C., Colomb, V., Decsi, T., Domellöf, M., et al., (2015). Arsenic in rice: A cause for concern. *J. Ped Gastroenterol Nutrit., 60*, 142–145.

Huang, W. X., Chen, X. W., Wu, L., Yu, Z. S., Gao, M. Y., Zhao, H. M., Mo, C. H., et al., (2021). Root cell wall chemistry remodeling enhanced arsenic fixation of a cabbage cultivar. *Journal of Hazardous Materials, 420*, 126165.

Hussain, M. M., Bibi, I., Niazi, N. K., Shahid, M., Iqbal, J., Shakoor, M. B., Ahmad, A., et al., (2021). Arsenic biogeochemical cycling in paddy soil-rice system: Interaction with various factors, amendments and mineral nutrients. *Science of the Total Environment, 773*, 145040.

Jobe, T. O., Yu, Q., Hauser, F., Xie, Q., Meng, Y., Maassen, T., Kopriva, S., & Schroeder, J. I., (2021). The SLIM1 transcription factor is required for arsenic resistance in *Arabidopsis thaliana*. *FEBS Letters, 595*, 1696–1707.

Keren, K., Krueger, M., Gilad, R., Ben-Yoseph, G., Sivan, U., & Braun, E., (2002). Sequence-specific molecular lithography on single DNA molecules. *Science, 297*, 72–75.

Khanam, R., Kulsum, P. G. P. S., Mandal, B., Hazra, G. C., & Kundu, D., (2022). The mechanistic pathways of arsenic transport in rice cultivars: Soil to mouth. *Environmental Research, 204*, 111942.

Kumarathilaka, P., Seneweera, S., Ok, Y. S., Meharg, A., & Bundschuh, J., (2019). Arsenic in cooked rice foods: Assessing health risks and mitigation options. *Environment International, 127*, 584–591.

Lafuente, A., Pajuelo, E., Caviedes, M. A., & Rodriguez-Llorente, I. D., (2010). Reduced nodulation in alfalfa induced by arsenic correlates with altered expression of early nodulins. *J. Plant Physiol, 167,* 286–291.

Li, W. X., Chen, T. B., Huang, Z. C., Lei, M., & Liao, X. Y., (2006). Effect of arsenic on chloroplast ultrastructure and calcium distribution in arsenic hyperaccumulator *Pteris vittata* L. *Chemosphere, 62*, 803–809.

Liu, W. J., Zhu, Y. G., Hu, Y., Williams, P. N., Gault, A. G., Meharg, A. A., Charnock, J. M., & Smith, F. A., (2006). Arsenic sequestration in iron plaque, its accumulation and speciation in mature rice plants (*Oryza sativa* L.). *Environmental Science & Technology, 40*(18), 5730–5736.

Liu, Y., Li, M., Han, C., Wu, F., Tu, B., & Yang, P., (2013). Comparative proteomic analysis of rice shoots exposed to high arsenate. *Journal of Integrative Plant Biology, 55*, 965–978.

Martinez, V. D., Vucic, E. A., Becker-Santos, D. D., Gil, L., & Lam, W. L., (2011). Arsenic exposure and the induction of human cancers. *J. Toxicol.*, 431287.

Martínez-Castillo, J. I., Saldaña-Robles, A., & Ozuna, C., (2022). Arsenic stress in plants: A metabolomic perspective. *Plant Stress, 3*, 100055.

Mawia, A. M., Hui, S., Zhou, L., Li, H., Tabassum, J., Lai, C., Wang, J., et al., (2021). Inorganic arsenic toxicity and alleviation strategies in rice. *Journal of Hazardous Materials, 408*, 124751.

Meng, X. Y., Qin, J., Wang, L. H., Duan, G. L., Sun, G. X., Wu, H. L., Chu, C. C., et al., (2011). Arsenic biotransformation and volatilization in transgenic rice. *New Phytol., 191*, 49–56.

Mishra, S., Stark, H. J., & Kupper, H., (2014). A different sequence of events than previously reported leads to arsenic-induced damage in *Ceratophyllum demersum* L. *Metallomics, 6*, 444–454.

Mishra, S., Tripathi, R. D., Srivastava, S., Dwivedi, S., Trivedi, P. K., Dhankher, O. P., & Khare, A., (2009). Thiol metabolism play a significant role during cadmium detoxification by *Ceratophyllum demersum* L. *Bioresource Technology, 100*, 2155–2161.

Moulick, D., Ghosh, D., Skalicky, M., Gharde, Y., Mazumder, M. K., Chowdhury, S., et al., (2022). Interrelationship among rice grain arsenic, micronutrients content and grain quality attributes: An investigation from genotype× environment perspective. *Frontiers in Environmental Science,* 180.

Moulick, D., Samanta, S., Sarkar, S., Mukherjee, A., Pattnaik, B. K., Saha, S., Awasthi, J. P., Bhowmick, S., Ghosh, D., et al., (2021). Arsenic contamination, impact and mitigation strategies in rice agro-environment: An inclusive insight. *Science of The Total Environment, 800*, 149477.

Muehe, E. M., Wang, T., Kerl, C. F., Planer-Friedrich, B., & Fendorf, S., (2019). Rice production threatened by coupled stresses of climate and soil arsenic. *Nature Communications, 10,* 1–10.

Mylona, P. V., Polidoros, A. N., & Scandalios, J. G., (1998). Modulation of antioxidant responses by arsenic in maize. *Free Radical Biology and Medicine, 25*, 576–585.

Norton, G., Lou-Hing, D. E., Meharg, A. A., & Price, A. H., (2008). Rice-arsenate interactions in hydroponics: Whole genome transcriptional analysis. *J. Exp. Bot., 59*, 2267–2276.

Pan, D., Yi, J., Li, F., Li, X., Liu, C., Wu, W., & Tao, T., (2020). Dynamics of gene expression associated with arsenic uptake and transport in rice during the whole growth period. *BMC Plant Biology, 20*(1), 1–10.

Pathare, V., Srivastava, S., & Suprasanna, P., (2013). Evaluation of effects of arsenic on carbon, nitrogen and sulfur metabolism in two contrasting varieties of *Brassica juncea. Acta Physiol. Plant., 35*, 3377–3389.

Punshon, T., Carey, A. M., Ricachenevsky, F. K., & Meharg, A. A., (2018). Elemental distribution in developing rice grains and the effect of flag-leaf arsenate exposure. *Environmental and Experimental Botany, 149*, 51–58.

Raessler, M., (2018). The arsenic contamination of drinking and groundwaters in Bangladesh: Featuring biogeochemical aspects and implications on public health. *Arch Environ. Contam Toxicol., 75*, 1–7. 10.1007/s00244-018-0511-4.

Requejo, R., & Tena, M., (2005). Proteome analysis of maize roots reveals that oxidative stress is a main contributing factor to plant arsenic toxicity. *Phytochemistry, 66*, 1519–1528.

Requejo, R., & Tena, M., (2006). Maize response to acute arsenic toxicity as revealed by proteome analysis of plant shoots. *Proteomics, 6*, S156–S162.

Rosen, B. P., Ajees, A. A., & McDermott, T. R., (2011). Life and death with arsenic. *Bioessays, 33*, 350–357.

Scheneider, J., Labory, C. R. G., Rangel, W. M., Alves, E., & Guilherme, L. R. G., (2013). Anatomy and ultrastructure alterations of *Leucaena leucocephala* (Lam.) inoculated with mycorrhizal fungi in response to arsenic-contaminated soil. *Journal of Hazardous Materials, 262*, 1245–1258.

Shaheen, S. M., Kwon, E. E., Biswas, J. K., Tack, F. M., Ok, Y. S., & Rinklebe, J., (2017). Arsenic, chromium, molybdenum, and selenium: Geochemical fractions and potential mobilization in riverine soil profiles originating from Germany and Egypt. *Chemosphere, 180*, 553–563.

Shahid, M. A., Balal, R. M., Khan, N., Zotarelli, L., Liu, G. D., Sarkhosh, A., et al., (2019). Selenium impedes cadmium and arsenic toxicity in potato by modulating carbohydrate and nitrogen metabolism. *Ecotoxicol. Environ. Saf., 180*, 588–599.

Shahid, M., Imran, M., Khalid, S., Murtaza, B., Niazi, N. K., Zhang, Y., & Hussain, I., (2020). Arsenic environmental contamination status in South Asia. In: *Arsenic in Drinking Water and Food* (pp. 13–39). Singapore: Springer.

Sharma, N., Tripathi, A., & Sanan-Mishra, N., (2015). Profiling the expression domains of a rice-specific miroRNA under stress. *Front Plant Sci., 6*, 333.

Shri, M., Kumar, S., Chakrabarty, D., Trivedi, P. K., Mallick, S., Misra, P., Shukla, D., et al., (2009). Effect of arsenic on growth, oxidative stress, and antioxidant system in rice seedlings. *Ecotoxicology and Environmental Safety, 72*, 1102–1110.

Singh, H. P., Batish, D. R., Kohli, R. K., & Arora, K., (2007). Arsenic-induced root growth inhibition in mung bean (*Phaseolus aureus* roxb.) is due to oxidative stress resulting from enhanced lipid peroxidation. *Plant Growth Regulation, 53*(1), 65–73.

Srivastava, A. K., Pandey, M., Ghate, T., Kumar, V., Upadhyay, M. K., Majumdar, A., Sanjukta, A. K., et al., (2021). Chemical intervention for enhancing growth and reducing grain arsenic accumulation in rice. *Environmental Pollution, 276*, 116719.

Srivastava, S., & D'Souza, S. F., (2009). Increasing sulfur supply enhances tolerance to arsenic and its accumulation in *Hydrilla verticillata* (L.f.) royle. *Environ. Sci. Technol., 43*, 6308–6313.

Srivastava, S., & D'Souza, S. F., (2010). Effect of variable sulfur supply on arsenic tolerance and antioxidant responses in *Hydrilla verticillata* (L.f.) royle. *Ecotox. Environ. Safety, 73*, 1314–1322.

Srivastava, S., Akkarakaran, J. J., Sounderajan, S., Shrivastava, M., & Suprasanna, P., (2016). Arsenic toxicity in rice (*Oryza sativa* L.) is influenced by sulfur supply: Impact on the expression of transporters and thiol metabolism. *Geoderma, 270*, 33–42.

Srivastava, S., Akkarakaran, J. J., Suprasanna, P., & D'Souza, S. F., (2013a). Response of adenine and pyridine metabolism during germination and early seedling growth under arsenic stress in *Brassica juncea*. *Acta Physiol. Plant., 35*, 1081–1091.

Srivastava, S., Mishra, S., Tripathi, R. D., Dwivedi, S., Trivedi, P. K., & Tandon, P. K., (2007). Phytochelatins and antioxidant systems respond differentially during arsenite and arsenate stress in *Hydrilla verticillata* (L.f.) royle. *Environ. Sci. Technol., 41*, 2930–2936.

Srivastava, S., Pathak, S., Ponsin, M., Hensawang, S., Chanpiwat, P., Yoeurn, C., & Phan, K., (2022). Sustainable solutions to arsenic accumulation in rice grown in south and south-east Asia. *Crop and Pasture Science, 73*, 149–159.

Srivastava, S., Pathare, V. S., Sounderajan, S., & Suprasanna, P., (2019). Nitrogen supply influences arsenic accumulation and stress responses of rice (*Oryza sativa* L.) seedlings. *Journal of Hazardous Materials, 367*, 599–606.

Srivastava, S., Srivastava, A. K., Singh, B., Suprasanna, P., & D'Souza, S. F., (2013b). The effect of arsenic on pigment composition and photosynthesis in *Hydrilla verticillata*. *Biol. Plant., 57*, 385–389.

Srivastava, S., Srivastava, A. K., Suprasanna, P., & D'souza, S. F., (2009). Comparative biochemical and transcriptional profiling of two contrasting varieties of *Brassica juncea* L. in response to arsenic exposure reveals mechanisms of stress perception and tolerance. *Journal of Experimental Botany, 60*(12), 3419–3431.

Srivastava, S., Srivastava, A. K., Suprasanna, P., & D'souza, S. F., (2013d). Identification and profiling of arsenic stress-induced microRNAs in *Brassica juncea*. *J. Exp. Bot., 64*, 303–315.

Srivastava, S., Srivastava, A. K., Suprasanna, P., & D'Souza, S. F., (2013c). Quantitative real-time expression profiling of aquaporin-isoforms and growth response of *Brassica juncea* under arsenite stress. *Mol. Biol. Rep., 40*, 2879–2886.

Srivastava, S., Suprasanna, P., & D'Souza, S. F., (2011). Redox state and energetic equilibrium determine the magnitude of stress in *Hydrilla verticillata* upon exposure to arsenate. *Protoplasma, 248*, 805–815.

Sundaram, S., Wu, S., Ma, L. Q., & Rathinasabapathi, B., (2009). Expression of a *Pteris vittata* glutaredoxin PvGRX5 in transgenic *Arabidopsis thaliana* increases plant arsenic tolerance and decreases arsenic accumulation in the leaves. *Plant Cell Environ., 32*, 851–858.

Tawfik, D. S., & Viola, R. E., (2011). Arsenate replacing phosphate: Alternative life chemistries and ion promiscuity. *Biochemistry, 50*, 1128–1134.

Thounaojam, T. C., Khan, Z., Meetei, T. T., Srivastava, S., Panda, S. K., & Upadhyaya, H., (2021). Transporters: The molecular drivers of arsenic stress tolerance in plants. *Journal of Plant Biochemistry and Biotechnology, 30*, 730–743.

Upadhyay, M. K., Majumdar, A., Srivastava, A. K., Bose, S., Suprasanna, P., & Srivastava, S., (2022). Antioxidant enzymes and transporter genes mediate arsenic stress reduction in rice (*Oryza sativa* L.) upon thiourea supplementation. *Chemosphere, 292*, 133482.

Wang, C., Zheng, L., Tang, Z., Sun, S., Ma, J. F., Huang, X. Y., & Zhao, F. J., (2020). OASTL-A1 functions as a cytosolic cysteine synthase and affects arsenic tolerance in rice. *Journal of Experimental Botany, 71*, 3678–3689.

Xu, J., Shi, S., Wang, L., Tang, Z., Lv, T., Zhu, X., Ding, X., et al., (2017). OsHAC4 is critical for arsenate tolerance and regulates arsenic accumulation in rice. *New Phytologist, 215*(3), 1090–1101.

Xue, X. M., Ye, J., Raber, G., Rosen, B. P., Francesconi, K., Xiong, C., Zhu, Z., et al., (2019). Identification of steps in the pathway of arsenosugar biosynthesis. *Environ. Sci. Technol., 53*, 634–641.

Yadav, P., & Srivastava, S., (2021). Effect of thiourea application on root, old leaf and young leaf of two contrasting rice varieties (*Oryza sativa* L.) grown in arsenic contaminated soil. *Environ. Techno. Innov., 21*, 101368.

Yadav, P., Srivastava, S., Patil, T., Raghuvanshi, R., Srivastava, A. K., & Suprasanna, P., (2021). Tracking the time-dependent and tissue-specific processes of arsenic accumulation and stress responses in rice (*Oryza sativa* L.). *J. Hazard Mater., 406*, 124307.

Ye, Y., Li, P., Xu, T., Zeng, L., Cheng, D., Yang, M., Luo, J., & Lian, X., (2017). OsPT4 contributes to arsenate uptake and transport in rice. *Frontiers in Plant Science, 8*, 2197.

Yu, L. J., Luo, Y. F., Liao, B., Xie, L. J., Chen, L., Xiao, S., et al., (2012). Comparative transcriptome analysis of transporters, phytohormone and lipid metabolism pathways in response to arsenic stress in rice (*Oryza sativa*). *New Phytol., 195*, 97–112.

Zhao, F. J., McGrath, S. P., & Meharg, A. A., (2010). Arsenic as a food chain contaminant: Mechanisms of plant uptake and metabolism and mitigation strategies. *Annu. Rev. Plant Biol., 61*, 535–559.

Zheng, M. Z., Li, G., Sun, G. X., Shim, H., & Cai, C., (2013). Differential toxicity and accumulation of inorganic and methylated arsenic in rice. *Plant Soil, 365*, 227–238.

CHAPTER 2

Status of Arsenic Contamination in Paddy Fields: Soil Water Chemistry

VINAY ASERI,[1] VARAD NAGAR,[2] ANUJ SHARMA,[2] APOORVA SINGH,[2] MAHIPAL SINGH SANKHLA,[3] and KUMUD KANT AWASTHI[4]

[1]*School of Forensic Science and Risk Management, Rashtriya Raksha University, Gandhinagar, Gujrat, India*

[2]*Department of Forensic Science, Vivekananda Global University, Jaipur India.*

[3]*Department of Forensic Science, University Centre for Research and Development (UCRD), Chandigarh University, Mohali, Punjab, India.*

[4]*Department of Life Sciences, Vivekananda Global University, Jaipur, India*

ABSTRACT

Rice is the major source of carbohydrates for billions of people throughout the world. Rice grains have significant arsenic (As) contents due to natural geogenic and human sources. Or As is highly bioavailable to rice roots under rice-growing circumstances, this is the case. To detect regional and temporal variations in As contamination levels in paddy soil-water systems, a multidimensional and inter-disciplinary knowledge of both short- and long-term consequences is required. During flooding, the inorganic content of soil pore fluids rises. Anaerobism results in weakly mobile As(V), which is converted to highly mobile As in dryland agricultural methods (III). The presence of iron (Fe) plaque on roots, metal (hydro)oxides (Fe and Mn), organic matter, clay mineralogy, and competing ions and compounds (PO_4

Arsenic in Rice: A Global Menace. Garima Awasthi, Sudhakar Shrivastava, and Mahipal Singh Sankhla, (Eds.)

and $Si(OH)_4$) in paddy soil-water systems are all known to impact As(V) and As(III) mobility. Microorganisms are important in As transformation via oxidation/reduction and methylation/volatilization processes, although the kinetics of these events are poorly understood. The improvement of all biogeochemical factors based on scientific evidence might assist in drastically minimizing inorganic As bioavailability.

2.1 INTRODUCTION

Arsenic (As) levels are found in many countries across the world, posing a threat to the environment. Inorganic is arranged as a meeting by the United States Environmental Protection Agency (USEPA). A cancer-causing substance Constant openness to inorganic substances across time Diseases of the skin, lungs, kidneys, liver, and bladder, as well as cardiovascular, neurological, hematological, renal, and respiratory ailments, affect humans. Furthermore, it is a contributing factor in diabetes mellitus, DNA, protein, and lipid damage, and has a negative impact on children's academic performance. As a result of the continuous toxicological effects of As, several experts' recommendations and administrative cutoff thresholds have been revised. Arsenic will generally accumulate in greater fixations in several foodstuffs, most notably rice, Oryza sativa L. The World Health Organization's (WHO) acceptable threshold for absolute As in drinking water was reduced from 50 to 10 mgl^{-1} in 1993, followed by the USEPA in 2001. Rice is the primary source of nutrition for over 3 billion people, the majority of whom live in Asia, supplying 70% and half of their calorie and protein needs, respectively. According to the United Nations, 115 countries grow rice, with 2e3 trimming seasons in roughly half of Asia's countries. African, Latin American, and Caribbean countries are limited to two seasons each year. Rice creation in all calm locals is limited to one season each year with top rice Production a very long time in each season contrasting between nations (Laborte et al., 2017).

Paddy rice is worked out how to ensure a standing head of water over the dirt and includes: (i) furrowing to homogenize waterlogged dirts; (ii) maintaining a standing head of water in the fields during the editing time frame; and (iii) water depleted and soil allowed to dry out before collection. To cope with the redox science of the paddy soil-water climate, flooding, and depletion are used. Arsenic accumulates in paddy soil-water by natural mechanisms (such as rock enduring and alluvial storage). The usage of As-depleted groundwater to inundate rice also results in the growth of As. Absolute As in water system water should be less than 100 mg L^{-1}, according to the Food and Agricultural Organization (FAO). Mining for Ag, Au, Cu, Pb, W, and Zn, as well as other associated activities, can contaminate nearby paddies with As. Phosphate (PO_4^{3+})-based composts, cow fertilizer, and the use of arsenical-combined insecticides in cotton-rice pivots are some of the several sources of As in paddy soils. In agricultural soils, the European Union (EU) has set a limit of 20 mg kg^{-1} for As. We discuss the temporal and spatial appropriations of As in paddy soil-water settings in relation to the physicochemical features of the dirt in this audit. The role of the microbiome in As transport and transition between distinct arsenic species in paddy soil water will also be investigated.

2.2 THE RELATIONSHIP BETWEEN ARSENIC SPECIATION AND PHYSIOCHEMICAL CHARACTERISTICS

2.2.1 pH AND Eh

In the climate, arsenic exists in a few oxidation states (for example, –3, 0, +3, and +5) and occurs in both strong and fluid phases; nevertheless, it is often found in nature in its local state, As(0) or As(–3) (Wang and Mulligan 2006). With a pH range of of 6.5-8.5, As will often activate quickly in groundwater. In paddy soil pore fluids, As is generally found in its inorganic structure as oxyanions of trivalent arsenite (As(III)) or pentavalent arsenate (As(V)), with monomethyl arsenic corrosive (MMA(V)) and dimethylarsinic corrosive (DMA(V)) being the most commonly recognized natural As species (Campbell, Nordstrom, and Geochemistry 2014) It was discovered that the pH of the soil is strongly linked to the concentration of complete As in rice grains. This is because a high pH causes negative surface charges, which might speed up the desorption of As(III) and As(V) into the soil. The redox potential (Eh) of As species determines their speciation, flexibility, and bioavailability. At low pH, arsenic corrosive (H_3AsO_4) is the most common animal kind under oxidizing circumstances (Wang and Mulligan 2006). In the paddy environment, many mechanisms, such as the reductive disintegration of metal hydro(oxides), the creation of root plaques, and microbial-interceded decline and oxidation, are for the most part contributors to As speciation (Smedley and Kinniburgh 2002).

2.2.2 ORGANIC MATTER

As factors in the paddy environment, natural carbon obtained from disintegrating plant and creature objects plays a key role (Erdogan & Tareq,

2014). The natural matter is regularly applied to soil as fertilizer in rice cultivation. After harvest, rice stubble and roots are frequently re-fused into paddy fields (Said-Pullicino et al. 2016). This method will encourage the formation of broken-down natural carbon (DOC) in paddy soils, which will be supplemented under overburdened situations. The natural matter has a strong affinity for As, leading to the creation of organo-As structures. (Paikaray, Banerjee, and Mukherji 2005; Radloff et al. 2007; Williams et al. 2011; Rahman et al. 2010) revealed that combining Lathyrus, vermicompost, and poultry manure into As contaminated paddy soils reduced absolute As concentration in rice tissues. Additionally, (Fu et al. 2011) observed a negative association between dirt natural matter levels and absolute As fixation in rice grains. In any event, DOC may be adsorbed onto Fe(III) (hydro)oxides (FeOOH) via ligand exchange, and so compete with both As(III) and As(V) for dynamic adsorption sites. As a result, DOC discovered in the soil configuration may improve the mobility and bioavailability of both As(III) and As(V) (Mladenov et al. 2015). In addition, DOC stimulates Fe(III)-degrading bacteria, accelerating microbial FeOOH reduction and triggering the entry of As(III) into the soil arrangement (Mladenov et al. 2015).

2.2.3 CLAY MINERALOGY

Dirt minerals like as kaolinite, montmorillonite, and illite are shaped mostly by the alteration of rocks due to physical, chemical, and organic factors (Regmi et al. 2013; Wimpenny et al. 2014). Improved flocculation and mud particle dispersion are caused by a lower Eh produced under overburdened circumstances. Mud minerals move to the lowest section of the furrowed layer in this manner, forming a furrow container (Li et al. 2009). Furrow container layout is also advanced by mechanical control by ranchers, animals, and heavy machinery. Plowing over a long period of time creates micropores (rather than mesopores and macropores) in the furrow container. As a result, the water maintenance limit in furrowed strata has been increased (Kögel-Knabner et al. 2010). Expanded water maintenance is a significant thought to limit water system water prerequisites, especially those with As pollution. The presence of dirt minerals prompts better-finished soils with a moderately high surface region. FeOOH is mostly consolidated in the dirt-size soil part and accordingly, mud soils might advance both As(III) and As(V) maintenance in paddy soils (Sahoo and Kim 2013). Surface spillover because of the weighty precipitation and the ensuing obliteration of soil structure through furrowing is answerable for the deficiency of mud minerals during paddy development and harms As immobilization (Kögel-Knabner et al. 2010).

2.2.4 IRON

Iron science has an important role in determining the mobility of As(III) and As(V) in paddy soils (Eqns. (1) and (2)) (Pan et al. 2014; Smedley and Kinniburgh 2002). FeOOH minerals such as ferrihydrite, goethite, lepidocrocite, and hematite may be found in paddy soils. Similarly, in overburdened situations, the arrangement of Fe plaque leads to the sequestration of As species (talked about exhaustively in segment 3.2). At the oxic/anoxic boundary of FeOOH, which has a large explicit surface region, both As(V) and As(III) can be adsorbed or co-hasten (for example, ferrihydrite: 100e700 m^2 g^1, goethite: 8e200 m^2 g^1, lepidocrocite: 15e260 m^2 g^1, hematite: 2e115 m^2 g^1). The sorption interaction is involved in the ligand exchange between inorganic As species and OH_2 and OH in Fe particle coordination rings. FeOOH reduces As(III) and (V) bioavailability in rice plants in this way (Youngran et al. 2007).

$$FeoH(s) + AsO_3^{3-} + 3 \rightarrow FeAsO_3 + H20$$

$$FeOH(s) + AsO_4^{3-} + 3H^{+} \rightarrow FeH_2AsO_4 + H20$$

Reductive disintegration of FeOOH creates As(III) adaptability in paddy soil-water frameworks, as mentioned previously (Huang et al. 2016; Yamaguchi et al. 2011). As a result, as compared to other grain crops, rice plants are destined to be prone to increase As accumulation. Agronomic methods, in general, cause modifications in Fe science in paddy soils, which impact As versatility and conveyance in a circuitous way.

2.2.5 MANGANESE

In paddy circumstances, manganese oxide minerals link with As speciation. Manganese oxides can degrade because they receive electrons throughout the oxidation cycle. In paddy soil water, As is versatile in two ways: by oxidizing As(III) to As(V) and by adsorbing As(V). In paddy soil water, As is versatile in two ways: by oxidizing As(III) to As(V) and by adsorbing As(V) (Liu et al. 2005). For example, one of the most notable multilayer Mn oxides, phyllo manganates, may quickly convert As(III) to As(V) (Manning et al., 2002, Tani et al. 2004). The ability of a few microorganisms to catalyze the manganese oxide acceptance of As(III) oxidation was demonstrated. Manganese plaque in rice roots also sequesters manganese. As a result, the absolute As in rice tissue and the Mn level in paddy soils typically have an inverse relationship (Hossain et al. 2008).

2.2.6 SULFUR

During overflowing circumstances, sulfate (SO_4^{2-}) in the soil network degrades into sulfide (S_2) as the redox potential drops (from 120 to 180 mV). In paddy soils, microbial SO_4^{2-} reduction also plays an important role in determining As portability (Jia et al. 2012, Wilkin, Wallschläger, and Ford 2003). Because As(III) precipitates when arsenic sulfide and iron sulfide minerals as sulfate levels drop to S_2, As(III) becomes immobilized (Burton et al. 2014; Fisher et al. 2008). Because arsenic sulfide and iron sulfide minerals precipitate when sulfate levels fall below S_2, As(III) gets immobilized (Burton et al. 2014). According to mackinawite arrangement and As(III) co-precipitation, microbial SO_4^{2-} reduction causes lower convergences of both Fe(II) and As(III) in spilled soil. Though As(III) sorption on iron sulfide is dependent on As speciation in watery medium and the mineralogy of iron sulfide (Isokpehi et al., 2012). Furthermore, As(III) can reply with S_2 and structure fluid thioarsenite structures under overloaded situations, which may stimulate As-S strong stages (i.e., undefined As_2S_3 or AsS) (Bostick, Fendorf, and Brown 2005) Ordinarily, thio arsenite edificted are framed under low Fe(II) and high natural matter conditions, not wit stand, thio arsenite buildings are didn't normally found in paddy soils.

2.3 ARSENIC DYNAMICS IN THE PADDY ENVIRONMENT

In a paddy climate, natural, and artificial factors contribute to As dispersion. Sums up the primary cause (such as As-tainted water system water, modern waste/wastewater releases, mining activities, and regular geogenic hotspots) for As presentation into the paddy climate and all out

As fixations in paddy soils in As-affected rice growth places around the world. The underlying factor that brings As into paddy settings is discussed in detail in the next subsections. At various stages of the rice growth cycle, illustrates essential features and reactions in As-sullied paddy soil-water frameworks.

2.3.1 EFFECT OF AS CONTAMINATED IRRIGATION WATER

Rice plant yield is largely determined by the type of water used in the irrigation system during the growth cycle. Around 150e200 mm of water is supplied to prepare the field for rice planting; however, in some cases, this quantity may reach up to 900 mm of water (Bouman and Tuong 2001). During the development season (three months), a volume of water of between 500 and 3,000 mm is required, and this amount fluctuates depending on climatic conditions, soil type, and rice genotypes (Abedin, Feldmann, and Meharg 2002). Water disasters are often high when dry conditions occur, and paddy lands become partially or completely dry. Rice development during dry years is dependent on the availability of surface water, such as waterways, lakes, dams, and groundwater sources. The majority of surface water sources are protected against pollution. As far as tainting is concerned (Sharma et al. 2014). Mineral lasting and anthropogenic actions, like as contemporary release and city-strong waste dump sites, may, nevertheless, introduce inorganic and natural As species into surface water (Li et al. 2009, Zheng et al. 2015). Due to a lack of openness and poor foundation construction, most ranchers in rice-producing areas are forced to rely on readily available groundwater supplies. As a result, contaminated groundwater is frequently used as water system water in many rice-producing countries. As a result, paddy soils can be successfully polluted with As over time, and in certain cases, the total As the level in the soil exceeds the recommended threshold for rural uses. As a result, most observers point to As collection in paddy soils with As-depleted water systems (Huang et al. 2016 and Rahman and Hasegawa 2011). Spring science is an important factor to consider when determining As fixations in groundwater, especially in As-affected spring locations. The depth of a well varies from one location to the next, and there is no evidence of a close association between As focuses in groundwater. Because of the huge disparities in precipitation and extensive depletion, groundwater levels will often fluctuate throughout the year. As a result, oxygen reacts with As-containing minerals (such as pyrite (FeS_2) and arsenopyrite (FeAsS)) to provide As (Fakhreddine et al. 2015 and Sun et al. 2016). When FeOOH levels fall below a certain threshold, As(III) is released into the groundwater (Xie et al. 2015). Iron(III)-depleting microorganisms also aid in the reduction of FeOH to give As(III). As a result of the combination of regular cycles and human interaction, inorganic As species are being introduced into rice-growing regions via the groundwater water system (Table 2.1).

Groundwater that has been flooded is usually channeled into rice fields (Kabir et al. 2016 and Polizzotto et al. 2013). When As contaminated groundwater is used for water supply, an increase in As concentration in paddy soil water is commonly seen. Close to the channel bay, high fixations constrict to typically low levels at great distances (Dittmar et al. 2007). Nonetheless, several studies have revealed that stream speed and As fixations in the water system water along the water system channel have a close association. Because of the extended home time, aggregation, and subsequent settling of As-bearing colloids occurs at a slower stream speed. As fixations seen throughout the water system channel (Dittmar et al. 2007 and Roberts et al. 2010). A faster stream speed, on the other hand, has a shorter home time, limiting the buildup and subsequent settling down of As-bearing colloids in the water system channels. As a result, at greater operating stream speeds, all out As fixations in water system water along the water system channel stay mostly unchanged (Dittmar et al. 2007; Roberts et al. 2010). The kind of water system channel (for example, a man-made large channel or a soil/earth ditch channel) has a beneficial impact on the total As concentration in water system water over the channel length (Neumann et al. 2014). Decreased all-around As fixations seen along water system conduits can also be caused by synthetic alterations. Most studies have focused on the total As concentration rather than the As speciation (As(III) and As(V)) in the water system water. Nonetheless, limited investigations revealed that As(III), which accounts for the majority of absolute As in groundwater, gradually decreases as it travels through water system channels. The logic for this is that the dispersion of barometrical O_2 into the water system water might cause As(III) to oxidize into As(V). Responsive intermediates (for example, Fe(IV) species) formed during the oxidation of Fe(II) by O_2 may also contribute to the oxidation of As(III) into As(V) (Neumann et al., 2014). Broken-up Fe(II) and Mn(II) are oxidized by dispersed ambient O_2 and then encouraged as

TABLE 2.1 The Primary Causes of As's Introduction to Paddy Settings and Total As Levels in Paddy Soils in As-Affected Rice-Growing Areas Across the World

Cause	Country	Number of Samples	Sampling Depth (cm)	Concentration	References
Groundwater irrigation	Bangladesh	113	0–60	4.5–68	Rahman et al. (2010)
–	–	56	0–15	10–70	Panaullah et al. (2009)
–	–	71	1–15	3.1–42.5	Meharg, & Rahman (2003)
–	–	224	0–15	11.41–61.04	Hossain et al. (2008)
–	–	–	0–1	149	Dittmar et al. (2007)
–	–	–	5–10	54	–
–	–	–	>30	4–11	–
Natural origin	India	–	0–15	2.41–11.09	Biswas, Biswas, & Santra (2014)
–	–	6	0–5	51.5	Srivastava, Jankov, Pogány, & Saxena (2011)
–	–	–	5–10	37.7	–
–	–	–	10–15	21.1	–
–	–	–	0–5	38.2	Norra et al. (2005)
–	–	–	50–60	20.4	–
–	–	–	100–110	11.1	–
–	–	–	10–15	4.6–9.72	Bhattacharya, Samal, Majumdar, & Santra (2010)
–	–	21	0–15	0.68–2.33	Piscaglia et al. (2012)
–	Cambodia	16	–	0.68–26.36	Phan et al. (2013)
–	–	23	0–20	0.8–18	Seyfferth, McCurdy, Schaefer, & Fendorf (2014)
–	Vietnam	16	–	6–20	Huang et al. (2016)

TABLE 2.1 *(Condinued)*

Cause	Country	Number of Samples	Sampling Depth (cm)	Concentration	References
–	Nepal	10	0–10	6.1–12.5	Dahal et al. (2008)
–	–	–	10–20	6.1–11.3	–
–	–	–	0–10	5.6–9.9	–
–	–	–	10–20	5.8–10.0	–
–	Taiwan	11	0–20	11.8–112	Hsu, Hsi, Huang, Liao, & Hseu (2012)
Surface water	Vietnam	2	–	9.5–11.0	Huang et al. (2016)
Wastewater	China	–	–	34	Zheng et al. (2015)
–	–	6	0–20	3.05–5.58	Liu et al. (2005)
Mining deposition	South Korea	135	0–30	2.97–172.07	Yun et al. (2016)
–	–	40	0–15	64.4	Kwon, Nejad, & Jung (2017)
–	China	27	–	9.14–25.84	Ma, Wang, Jia, & Yang (2017)
–	–	19	0–15	97–709	Liu et al. (2005)
–	–	–	0–20	2.5–19.2	Jiang et al. (2014)
–	–	–	–	0.1–45.9	Fu et al. (2011)
–	Bangladesh	15	0–15	14.83–41.97	Halim, Majumder, & Zaman (2015)
–	India	–	0–100	13.9–24.1	Halim et al. (2015)
–	–	–	–	50–90	Hundal, Singh, Singh, & Kumar (2013)
–	France	23	0–10	5–10	Williams et al. (2007)
–	Spain	25	0–10	4–11	–
–	USA	6	0–10	4–7	–

Fe(OH)3 and MnO_2 as the water system water passes down the channel (Dittmar et al., 2007). As(V) complexation and co-precipitation with metal hydro-oxides (such as Fe and Mn) reduce the total As fixations detected in water system channels (Figure 2.1).

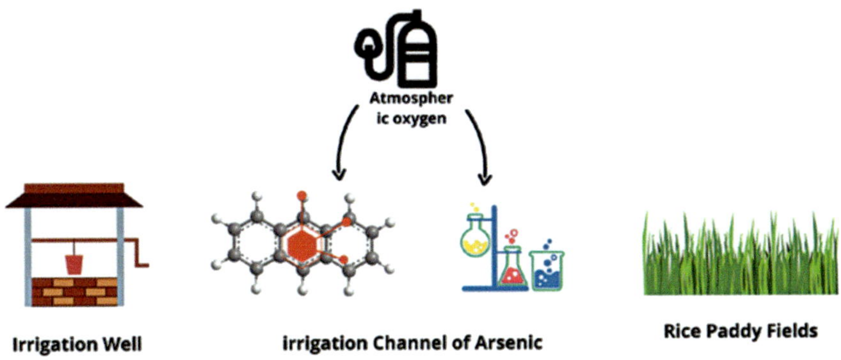

FIGURE 2.1 Variation of As and another elemental along the irrigation channel.

2.3.2 EFFECT OF MINING ACTIVITIES

Assuming that mining activities are active near rice fields, paddy soils might be polluted by As. Until now, a limited number of research have examined the effects of mining on As contamination in paddy soils, but no information on As speciation elements in mining-affected paddy soils during the rice plant's growth phase has been revealed. Mining of As-bearing minerals (such as arsenopyrite, pyrite, realgar (As), and orpiment (As_2S_3)), as well as optional As-bearing minerals (such as arsenopyrite, pyrite, realgar (As), and orpiment (As_2S_3) As minerals are to blame for the contamination of paddy soils with As. High precipitation and solid winds are to blame for tailings and waste rocks, caustic mining seepage, and wet and dry residue statements. As if accumulating in rice fields near mining areas (Kwon, Nejad, and Jung 2017; Yun et al. 2016), China produces the most rice in the world and is the finest example of mining-related As in paddy fields. Long-distance mining operations in Hunan, one of China's most important rice-producing regions, for example, resulted in significant all-out As fixations in adjacent paddy soils. Because of diverse parent materials, the example of full As appropriation in rice soils by mining activities fluctuates (Patel et al. 2005).

Absolute The concentrations in paddy soils on Hainan Island, China, varied between 0.1 and 45.9 mg kg^{-1}, depending on the stone types (transformative rocks: 18.9-45.9 mg kg^{-1}, corrosive molten rocks, and basalt: 0.1-5.9 and 0.2-4.2 mg kg^{-1}, clastic residue: 0.4-10.9 mg kg^{-1}) (Fu et al. 2011). In any case, all As focuses in mining-affected paddy soils are insufficient to provide meaningful data on As mobility and bioavailability. As a result, it's likely that the relevant natural threats will be addressed, with a focus on speciation levels. Synthetic fractionation using several extraction procedures provides detailed information on As in paddy soils (Fu et al. 2011, Hsu et al. 2012 and Norra et al. 2005). Paddy soils exposed to mining can have a greater amount of lingering stage As than the effectively versatile stage and redox-sensitive phases of As. Arsenic restriction to the next phases must be accomplished by significant ecological modifications (for example solid acids). The lingering bound component is more stable than the inexactly bound and redox delicate stages in this regard. However, paddy soils contaminated by As-polluted water system water have a different narrative to tell about substance fractionation. Those dirt have a bigger convergence of As held in inexactly bound phases and redox-touchy stages, whereas lingering stages have definitely the least grouping of As. Because of its co-precipitation with or sorption onto insufficiently glasslike metal hydro(oxides) or other components, such as transparent metal hydro(oxides), earth minerals, and natural matter of the dirt lattice, As is gathered in paddy soils owing to As rich groundwater. Surprisingly, paddy soils contaminated by mining chores contain a lot of residual bound As, which reduces As's mobility and bioavailability.

2.4 ROLE OF MICROORGANISMS ON AS DYNAMICS IN PADDY SOIL-WATER SYSTEM

In paddy soils, anaerobic and aerobic bacteria are related to As speciation via several cycles. (Zhang et al. 2015). Figure 2.2 Microorganisms in paddy soil water frameworks represent some different immediate As change paths.

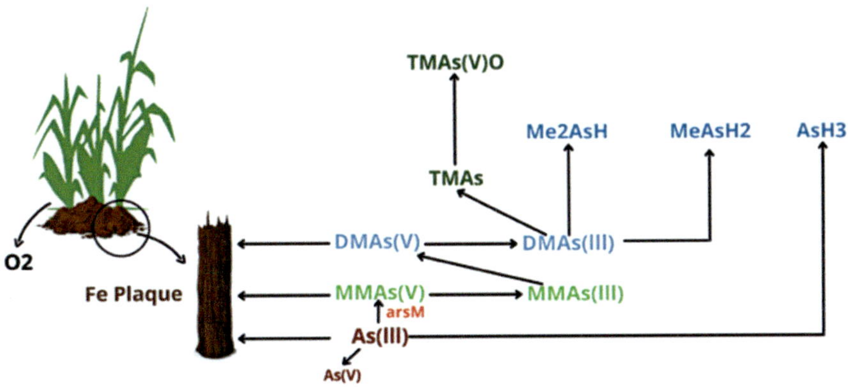

FIGURE 2.2 Microbiologically mediated in the paddy soil-water environment, As methylation and volatilization, As(III) oxidation, As(V) reduction, and dissimilative reduction have all been seen.

2.4.1 ARSENIC OXIDATION AND REDUCTION

Both As oxidation and decrease are attributed to microorganisms found in the rhizosphere and bulk paddy soil. For As transformation catalysts and transporters, a variety of properties found in microorganisms are encoded (Zecchin, Corsini, Martin, & Cavalca, 2017). Microorganisms' detoxification of As(III) to As(V), which is less toxic and portable than As(III), is coded for by the aioA quality code (Zhang et al., 2015). In paddy soils, microbial oxidation of As(III) occurs in both flooded and non-flooded environments. Paracoccus species may oxidize As(III) in paddy soil under both aerobic and anaerobic conditions by using O_2 and NO_3 as independent electron acceptors (Norra et al. 2005). In addition, As(III) serves as an electron donor during bacterial digestion (Zheng et al. 2015). Microbiologically-interceded As(III) oxidation becomes common in non-overflowed paddy soil once abiotic oxidation of As(III) stops due to passivation of receptive locales on mineral surfaces, despite the fact that the rate of transformation of As(III) to As(V) is a sluggish cycle, around 2.5e2.8% per day. Plant development-advancing (PGP) substances are improved by As(III) oxidizing microbes. For example, a gram-positive bacterium, Bacillus flexus ASO-6, isolated from the rhizosphere of rice paddy plants showed increased rates of As(III) oxidation (Vmax 14.34 mM min^1 107 cells) and released plant development-promoting compounds such as siderophores, IAA, and ACC deaminase (Das, Jean, Chou, Rathod,

& Liu, 2016). In this way, vaccination of As(III) oxidizing microorganisms is beneficial for limiting As weight and advancing rice plant growth. In paddy soils, microorganisms may reduce As(V) to As(III), but only under overflowing circumstances. As(V) reduction is linked to two different processes, namely As detoxification and the respiratory route, which are coded for individually by the arsC and arrA traits (Malasarn et al., 2004; Villegas-Torres et al., 2011). Under non-overflowed circumstances, phylogenetically diverse bacteria may considerably reduce As(V) to As(III) (Bachate, Cavalca, & Andreoni, 2009). Microbial reduction of As(V) to a more adaptable form of the paddy environment, As(III) may increase As mobility and bioavailability under both overloaded and non-overflowed situations. Apart from direct As alterations, microorganisms may be able to monitor As speciation in paddy soil-water frameworks in a roundabout way. Microscopic organisms that deplete Fe(III) and oxidize Fe(II) play a crucial role in As alteration (Zecchin et al. 2017). In persistently overburdened circumstances, the total overflow of Fe(III)-reducing bacteria is substantial. Under flooded circumstances, the reduction of Fe(III) minerals and subsequent entry of As(III) is expected to be substantial (Ohtsuka et al., 2013). The use of soil modifications, such as biochar, may have an influence on As(V) lowering (Wang & Mulligan, 2006). The biochar can stimulate the growth and overflow of Fe(III), reducing bacterial proliferation (for example, Clostridium, Bacillus, and Caloramator) and allowing As(V) to operate more effectively (Wang & Mulligan, 2006). Poor (low) centralizations of Fe(II) and substantial degrees of Fe(III) may also reduce the activity and overflow of Fe(III) in paddy soils, reducing microscopic organisms. Despite this, little is known about the microbial community that is responsible for Fe(II) oxidation. In overflowing paddy soils, nitrate-subordinate Fe(II)-oxidizing microorganisms can oxidize Fe(II) to Fe (III). In such situations, arsenic species may promote or adsorb onto Fe(III) minerals in root plaque and bulk soils (Dong et al. 2014). As previously stated in Section 2.4.3, SO_4^{2-} depleting and SO_4^{2-} oxidizing microorganisms may also have an influence. In paddy soil-water frameworks, as speciation. In general, water board practices (such as overflowed and non-flooded situations), as well as physicochemical and ecological parameters, have a significant role in determining the overflow and diversity of microorganisms in paddy soil frameworks (Zecchin et al. 2017). Under overloaded conditions, for example, the total overflow of As-, Fe-, and Sreducing microorganisms expands, but under non-overflowed conditions, the converse is true (Das et al. 2016 and Zecchin et al. 2017). However, no definitive data on microbial interceded As oxidation/decrease energy and related rate restriction factors that may also dial down the reactions exists.

2.5 CONCLUSION AND FUTURE OUTREACH

The spatial distribution of As species in paddy soil-water exhibits significant heterogeneity, both horizontally and vertically, and it also varies over time. Relying solely on the total As concentration in paddy soil-water from a single location is insufficient. Hence, it is crucial to accurately comprehend the spatial and temporal variations of As species. There are evident gaps in our current understanding, which necessitate further exploration in future research. Additionally, special attention should be given to the impact of natural geogenic sources, such as weathering of As-bearing minerals and alluvial deposits, on the As loadings in paddy soil. It is imperative to conduct in-depth investigations into the soil mineralogy in regions where rice cultivation takes place. The presence of As-contaminated rice fields might lead to a decrease in the uptake of As species by rice roots. However, it is essential to be aware of potential secondary contamination issues associated with the mentioned amendments. One possible approach to mitigate As toxicity in paddy environments is through biostimulation, which involves providing nutrients that promote microbial growth. This, in turn, can enhance microbial-mediated processes such as As oxidation, methylation, and volatilization, leading to reduced As levels. Another viable option is to focus on breeding and introducing rice cultivars that can decrease the accumulation of As in the grain. Exploring this avenue of research could yield promising results in tackling the issue of As contamination in rice fields.

KEYWORDS

- **Rice**
- **Arsenic Chemistry**
- **Heavy Metal**
- **Arsenic accumulation**

REFERENCES

Abedin, M.J., Cresser, M.S., Meharg, A.A., Feldmann, J., Cotter-Howells, J., 2002a. Arsenic accumulation and metabolism in rice (Oryza sativa L.). *Environ. Sci. Technol. 36* (5), 962–968.

Bachate, S., Cavalca, L., Andreoni, V., 2009. Arsenic-resistant bacteria isolated from 584 agricultural soils of Bangladesh and characterization of arsenate-reducing strains. *J. Appl. 585 Microbiol. 107* (1), 145–156.

Bhattacharya, P., Samal, A., Majumdar, J., Santra, S., 2009. Transfer of arsenic from 587 groundwater and paddy soil to rice plant (Oryza sativa L.): a micro level study in West 588 Bengal, India. *World J. Agri. Sci. 5* (4), 425–431.

Biswas, A., Biswas, S., Santra, S.C., 2014. Arsenic in irrigated water, soil, and rice: perspective 598 of the cropping seasons. *Paddy Water Environ. 12* (4), 407–412.

Bostick, B., Fendorf, S., Brown, G., 2005. In situ analysis of thioarsenite complexes in neutral to alkaline arsenic sulphide solutions. *Mineral. Mag. 69* (5), 781–795.

Bouman, B., Tuong, T.P., 2001. Field water management to save water and increase its productivity in irrigated lowland rice. *Agri. Water Manage. 49* (1), 11–30.

Burton, E.D., Johnston, S.G., Kocar, B.D., 2014. Arsenic mobility during flooding of contaminated soil: the effect of microbial sulfate reduction. *Environ. Sci. Technol. 48* (23), 13660–13667.

Campbell, K.M., Nordstrom, D.K., 2014. Arsenic speciation and sorption in natural environments. *Rev. Mineral. Geochem. 79* (1), 185–216

Dahal, B.M., Fuerhacker, M., Mentler, A., Karki, K.B., Shrestha, R.R., Blum, W.E.H., 2008. 630 Arsenic contamination of soils and agricultural plants through irrigation water in *Nepal. 631 Environ. Pollut. 155* (1), 157–163.

Das, S., Chou, M.-L., Jean, J.-S., Liu, C.-C., Yang, H.-J., 2016a. Water management impacts 633 on arsenic behavior and rhizosphere bacterial communities and activities in a rice agro 634 ecosystem. *Sci. Total Environ. 542,* 642–652.

Dittmar, J., Voegelin, A., Roberts, L.C., Hug, S.J., Saha, G.C., Ali, M.A., Badruzzaman, 640 A.B.M., Kretzschmar, R., 2007. Spatial distribution and temporal variability of arsenic in 641 irrigated rice fields in Bangladesh. 2. Paddy soil. *Environ. Sci. Technol. 41* (17), 5967–5972.

Fakhreddine, S., Dittmar, J., Phipps, D., Dadakis, J., Fendorf, S., 2015. Geochemical triggers of arsenic mobilization during managed aquifer recharge. *Environ. Sci. Technol. 49* (13), 7802–7809.

Fu, Y., Chen, M., Bi, X., He, Y., Ren, L., Xiang, W., Qiao, S., Yan, S., Li, Z., Ma, Z., 2011. Occurrence of arsenic in brown rice and its relationship to soil properties from Hainan Island, China. *Environ. Pollut. 159* (7), 1757–1762.

Halim, M.A., Majumder, R.K., Zaman, M.N., 2015. Paddy soil heavy metal contamination and 669 uptake in rice plants from the adjacent area of Barapukuria coal mine, northwest 670 Bangladesh. *Arab. J. Geosci. 8* (6), 3391–3401.

Hossain, M., Jahiruddin, M., Panaullah, G., Loeppert, R., Islam, M., Duxbury, J., 2008. Spatial variability of arsenic concentration in soils and plants, and its relationship with iron, manganese and phosphorus. *Environ. Pollut. 156* (3), 739–744.

Hossain, M., Jahiruddin, M., Panaullah, G., Loeppert, R., Islam, M., Duxbury, J., 2008. Spatial 672 variability of arsenic concentration in soils and plants, and its relationship with iron, 673 manganese and phosphorus. *Environ. Pollut. 156* (3), 739–744.

Hsu, W.-M., Hsi, H.-C., Huang, Y.-T., Liao, C.-S., Hseu, Z.-Y., 2012. Partitioning of arsenic 675 in soil–crop systems irrigated using groundwater: a case study of rice paddy soils in 676 southwestern Taiwan. *Chemosphere 86* (6), 606–613.

Huang, Y., Miyauchi, K., Endo, G., Don, L.D., Manh, N.C., Inoue, C., 2016. Arsenic contamination of groundwater and agricultural soil irrigated with the groundwater in Mekong Delta, Vietnam. *Environ. Earth Sci. 75* (9), 757.

Huang, Y., Miyauchi, K., Endo, G., Don, L.D., Manh, N.C., Inoue, C., 2016. Arsenic contamination of groundwater and agricultural soil irrigated with the groundwater in Mekong Delta, Vietnam. *Environ. Earth Sci. 75* (9), 757.

Hundal, H.S., Singh, K., Singh, D., Kumar, R., 2013. Arsenic mobilization in alluvial soils of 684 Punjab, North–West India under flood irrigation practices. *Environ. Earth Sci. 69* (5), 1637– 685 1648.

Jiang, W., Hou, Q., Yang, Z., Zhong, C., Zheng, G., Yang, Z., Li, J., 2014. Evaluation of 697 potential effects of soil available phosphorus on soil arsenic availability and paddy rice 698 inorganic arsenic content. *Environ. Pollut. 188,* 159–165.

Kabir, M., Salam, M., Paul, D., Hossain, M., Rahman, N., Aziz, A., Latif, M., 2016. Spatial 705 Variation of Arsenic in Soil, Irrigation Water, and Plant Parts: A Microlevel Study. *Sci. 706 World J. 2016.*

Kögel-Knabner, I., Amelung, W., Cao, Z., Fiedler, S., Frenzel, P., Jahn, R., Kalbitz, K., Kölbl, A., Schloter, M., 2010. Biogeochemistry of paddy soils. *Geoderma 157* (1), 1–14.

Kwon, J.C., Nejad, Z.D., Jung, M.C., 2017. Arsenic and heavy metals in paddy soil and 712 polished rice contaminated by mining activities in Korea. *Catena 148,* 92–100.

Kwon, J.C., Nejad, Z.D., Jung, M.C., 2017. Arsenic and heavy metals in paddy soil and 712 polished rice contaminated by mining activities in Korea. *Catena 148,* 92–100.

Laborte, A.G., Gutierrez, M.A., Balanza, J.G., Saito, K., Zwart, S.J., Boschetti, M., Murty, M.,Villano, L., Aunario, J.K., Reinke, R., 2017. RiceAtlas, a spatial database of global rice calendars and production. *Sci. data 4.*

Li, R., Stroud, J., Ma, J., McGrath, S., Zhao, F., 2009. Mitigation of arsenic accumulation in rice with water management and silicon fertilization. Environ. *Sci. Technol. 43* (10), 3778– 721 3783.

Liu, H., Probst, A., Liao, B., 2005a. Metal contamination of soils and crops affected by the Chenzhou lead/zinc mine spill (Hunan, China). *Sci. Total Environ. 339 (1),* 153–166.

Ma, L., Wang, L., Jia, Y., Yang, Z., 2017. Accumulation, translocation and conversion of six 743 arsenic species in rice plants grown near a mine impacted city. *Chemosphere 183,* 44–52

Malasarn, D., Saltikov, C., Campbell, K., Santini, J., Hering, J., Newman, D., 2004. arrA is a 745 reliable marker for As (V) respiration. *Science 306* (5695), 455–455.

Manning, B.A., Fendorf, S.E., Bostick, B., Suarez, D.L., 2002. Arsenic (III) oxidation and arsenic (V) adsorption reactions on synthetic birnessite. *Environ. Sci. Technol. 36* (5), 976–981.

Meharg, A.A., Rahman, M.M., 2003. Arsenic contamination of Bangladesh paddy field soils: 753 implications for rice contribution to arsenic consumption. *Environ. Sci. Technol. 37* (2), 754 229–234.

Mladenov, N., Zheng, Y., Simone, B., Bilinski, T.M., McKnight, D.M., Nemergut, D., Radloff, K.A., Rahman, M.M., Ahmed, K.M., 2015. Dissolved organic matter quality in a shallow aquifer of Bangladesh: implications for arsenic mobility. *Environ. Sci. Technol. 49* (18), 10815–10824.

Neumann, R.B., Pracht, L.E., Polizzotto, M.L., Badruzzaman, A.B.M., Ali, M.A., 2014. 776 Sealing rice field boundaries in Bangladesh: a pilot study demonstrating reductions in water 777 use, arsenic loading to field soils, and methane emissions from irrigation water. Environ. 778 *Sci. Technol. 48* (16), 9632–9640.

Norra, S., Berner, Z., Agarwala, P., Wagner, F., Chandrasekharam, D., Stüben, D., 2005. 780 Impact of irrigation with As rich groundwater on soil and crops: a geochemical case study 781 in West Bengal Delta Plain, India. *Appl. Geochem. 20* (10), 1890–1906.

Norra, S., Berner, Z., Agarwala, P., Wagner, F., Chandrasekharam, D., Stüben, D., 2005. 780 Impact of irrigation with As rich groundwater on soil and crops: a geochemical case study 781 in West Bengal Delta Plain, India. *Appl. Geochem. 20* (10), 1890–1906.

Paikaray, S., Banerjee, S., Mukherji, S., 2005. Sorption of arsenic onto Vindhyan shales: Role of pyrite and organic carbon. *Curr. Sci.*, 1580–1585.

Pan, Y., Koopmans, G.F., Bonten, L.T.C., Song, J., Luo, Y., Temminghoff, E.J.M., Comans, R.N.J., 2014. Influence of pH on the redox chemistry of metal (hydr)oxides and organic matter in paddy soils. *J. Soils Sediment. 14* (10), 1713–1726.

Panaullah, G.M., Alam, T., Hossain, M.B., Loeppert, R.H., Lauren, J.G., Meisner, C.A., 792 Ahmed, Z.U., Duxbury, J.M., 2009. Arsenic toxicity to rice (Oryza sativa L.) in Bangladesh. *793 Plant Soil 317* (1–2), 31

Patel, K., Shrivas, K., Brandt, R., Jakubowski, N., Corns, W., Hoffmann, P., 2005. Arsenic 795 contamination in water, soil, sediment and rice of central India. *Environ. Geochem. Hlth. 796 27* (2), 131–145.

Phan, K., Sthiannopkao, S., Heng, S., Phan, S., Huoy, L., Wong, M.H., Kim, K.-W., 2013. 798 Arsenic contamination in the food chain and its risk assessment of populations residing in 799 the Mekong River basin of Cambodia. *J. Hazard. Mater. 262,* 1064–1071.

Polizzotto, M.L., Lineberger, E.M., Matteson, A.R., Neumann, R.B., Badruzzaman, A.B.M., 801 Ali, M.A., 2013. Arsenic transport in irrigation water across rice-field soils in Bangladesh. *802 Environ. Pollut. 179,* 210–217.

Radloff, K.A., Cheng, Z., Rahman, M.W., Ahmed, K.M., Mailloux, B.J., Juhl, A.R., Schlosser, P., van Geen, A., 2007. Mobilization of arsenic during one-year incubations of grey aquifer 805 sands from Araihazar, Bangladesh. *Environ. Sci. Technol. 41* (10), 3639.

Rahman, M.A., Hasegawa, H., 2011. High levels of inorganic arsenic in rice in areas where arsenic-contaminated water is used for irrigation and cooking. *Sci. Total Environ. 409* (22), 4645–4655.

Rahman, M.S., Abdul Mazid Miah, M., Khaled, H.M., Islam, A., Panaullah, G., 2010. Arsenic concentrations in groundwater, soils, and irrigated rice in Southwestern Bangladesh. Commun. *Soil Sci. Plan. 41* (16), 1889–1895.

Roberts, L.C., Hug, S.J., Dittmar, J., Voegelin, A., Kretzschmar, R., Wehrli, B., Cirpka, O.A., 819 Saha, G.C., Ali, M.A., Badruzzaman, A.B.M., 2010. Arsenic release from paddy soils 820 during monsoon flooding. *Nat. Geosci. 3* (1), 53.

Sahoo, P.K., Kim, K., 2013. A review of the arsenic concentration in paddy rice from the perspective of geoscience. *Geosci. J. 17* (1), 107–122.

Said-Pullicino, D., Miniotti, E.F., Sodano, M., Bertora, C., Lerda, C., Chiaradia, E.A., Romani, M., de Maria, S.C., Sacco, D., Celi, L., 2016. Linking dissolved organic carbon cycling to organic carbon fluxes in rice paddies under different water management practices. *Plant Soil 401* (1–2), 273–290.

Seyfferth, A.L., McCurdy, S., Schaefer, M.V., Fendorf, S., 2014. Arsenic concentrations in 838 paddy soil and rice and health implications for major rice-growing regions of Cambodia. 839 Environ. *Sci. Technol. 48* (9), 4699–4706.

Sharma, A.K., Tjell, J.C., Sloth, J.J., Holm, P.E., 2014. Review of arsenic contamination, exposure through water and food and low cost mitigation options for rural areas. *Appl. Geochem. 41,* 11–33.

Shrivastava, A., Barla, A., Singh, S., Mandraha, S., Bose, S., 2017. Arsenic contamination in 844 agricultural soils of Bengal deltaic region of West Bengal and its higher assimilation in 845 monsoon rice. *J. Hazard. Mater. 324,* 526–534.

Smedley, P., Kinniburgh, D., 2002. A review of the source, behaviour and distribution of arsenic in natural waters. *Appl. Geochem. 17* (5), 517–568.

Sun, J., Quicksall, A.N., Chillrud, S.N., Mailloux, B.J., Bostick, B.C., 2016. Arsenic mobilization from sediments in microcosms under sulfate reduction. *Chemosphere 153,* 254–261.

Tani, Y., Miyata, N., Ohashi, M., Ohnuki, T., Seyama, H., Iwahori, K., Soma, M., 2004. Interaction of inorganic arsenic with biogenic manganese oxide produced by a Mn-oxidizing fungus, strain KR21-2. *Environ. Sci. Technol. 38* (24), 6618–6624.

Wang, S., Mulligan, C.N., 2006. Occurrence of arsenic contamination in Canada: sources, behavior and distribution. *Sci. Total Environ. 366* (2–3), 701–721.

Wilkin, R.T., Wallschläger, D., Ford, R.G., 2003. Speciation of arsenic in sulfidic waters. *Geochem. T. 4* (1), 1–7.

Williams, P.N., Villada, A., Deacon, C., Raab, A., Figuerola, J., Green, A.J., Feldmann, J., 897 Meharg, A.A., 2007. Greatly enhanced arsenic shoot assimilation in rice leads to elevated 898 grain levels compared to wheat and barley. *Environ. Sci. Technol. 41* (19), 6854–6859.

Williams, P.N., Zhang, H., Davison, W., Meharg, A.A., Hossain, M., Norton, G.J., Brammer, H., Islam, M.R., 2011. Organic matter - solid phase interactions are critical for predicting arsenic release and plant uptake in Bangladesh paddy soils. *Environ. Sci. Technol. 45* (14), 6080–6087.

Xie, X., Wang, Y., Li, J., Yu, Q., Wu, Y., Su, C., Duan, M., 2015. Effect of irrigation on Fe (III)–SO 4 2− redox cycling and arsenic mobilization in shallow groundwater from the Datong basin, China: Evidence from hydrochemical monitoring and modeling. *J. Hydrol. 523,* 128–138.

Yamaguchi, N., Nakamura, T., Dong, D., Takahashi, Y., Amachi, S., Makino, T., 2011. Arsenic release from flooded paddy soils is influenced by speciation, Eh, pH, and iron dissolution. *Chemosphere 83* (7), 925–932.

Youngran, J., Maohong, F., Van Leeuwen, J., Belczyk, J.F., 2007. Effect of competing solutes on arsenic (V) adsorption using iron and aluminum oxides. *J. Environ. Sci. 19* (8), 910–919.

Yun, S.-W., Baveye, P.C., Kim, K.-B., Kang, D.-H., Lee, S.-Y., Son, J., Kim, D.-H., Yoon, Y.- 916 C., Yu, C., 2016a. Effect of postmining land use on the spatial distribution of metal (loid) s 917 and their transport in agricultural soils: Analysis of a case study of Chungyang, South Korea. *918 J. Geochem. Explor. 170,* 157–166.

Zecchin, S., Corsini, A., Martin, M., Cavalca, L., 2017a. Influence of water management on 923 the active root-associated microbiota involved in arsenic, iron, and sulfur cycles in rice 924 paddies. *Appl. Microbiol. Biot. 101* (17), 6725–6738.

Zheng, R., Chen, Z., Cai, C., Tie, B., Liu, X., Reid, B.J., Huang, Q., Lei, M., Sun, G., Baltrėnaitė, E., 2015. Mitigating heavy metal accumulation into rice (Oryza sativa L.) using biochar amendment - a field experiment in Hunan, China. Environ. *Sci. Pollut. R. 22 (14),* 11097–11108.

Zheng, R., Chen, Z., Cai, C., Tie, B., Liu, X., Reid, B.J., Huang, Q., Lei, M., Sun, G., 938 Baltrėnaitė, E., 2015. Mitigating heavy metal accumulation into rice (Oryza sativa L.) using biochar amendment - a field experiment in Hunan, China. *Environ. Sci. Pollut. R. 22* (14), 940 11097–11108.

CHAPTER 3

Rhizospheric Physiology of Rice Plants Towards Arsenic

ANSHIKA KAUSHIK,[1] ANJALI AWASTHI,[2] GARIMA AWASTHI,[3] and SUNIL K. GUPTA[4]

[1]*Department of Anthropology, University of Delhi, India*

[2]*Department of Zoology, University of Rajasthan, India*

[3]*Department of Life Sciences, Vivekananda Global University, India*

[4]*Key Laboratory of Tropical Forest Ecology, Xishuangbanna Tropical Botanical Garden, Chinese Academy of Sciences, China*

ABSTRACT

Rice is a staple food and a high quantity of arsenic is found to be present in rice than other cereal crops. Arsenic exists in inorganic and organic forms. Different species of arsenic are transported to the plants via various transporters such as phosphate transporters, silicone transporters, aquaporin channels, etc. In rice cultivation, the iron plaque, leads to changes in the concentration of vital components such as manganese, iron, soil pH, microbial activities, soil-pH, soil organic matter, etc., leading to the environments that favor the preponderance of mobile arsenite As(III). Arsenic speciation, genetic variation, organic matter, and soil pH are the factors that affect the translocation of arsenic. This poses a great threat to public health as rice is a widely consumed food across the country.

Arsenic in Rice: A Global Menace. Garima Awasthi, Sudhakar Shrivastava, and Mahipal Singh Sankhla, (Eds.)

3.1 INTRODUCTION

Rhizosphere is a result of plant-soil interaction. It can be defined as the area of the root, which is populated by an inimitable group of microbes, attracted to the substances released by the root of the plant. It is also known as a microbe storehouse (Kennedyand, 2005). The wastes of the root system, either organic or inorganic, are known as root exudates, which influenced the rate of microbial activity in the rhizosphere. Rhizosphere can be divided into three distinct zones grounded on the propinquity and influence with respect to the root. The three zones are Endo rhizosphere, rhizoplane, and ecto-rhizosphere. The endo-rhizosphere embraces part of the cortex along with endodermis where microorganisms and cations can inhabit the "free space" amongst cells that are known as apoplastic space. The medial zone after the root comprising the root epidermis and mucilage is known as rhizoplane. The outermost zone of the rhizosphere, i.e., ecto-rhizosphere extends from the rhizoplane out into most of the surrounding soil (**McNear, 2013).**

3.2 ROOT STRUCTURE OF RICE

The roots of rice plants have a fibrous root system, consisting of three types of roots – nodal root, ephemeral root, and lateral roots (Figure 3.1). The root that originated whence the embryo is called the ephemeral seminal root and lives for about 30 days only. The roots which originate via the basal internode of the stem part are known as nodal roots. Lateral roots are the roots that emerge from other roots and can be identified as thick lateral roots and fine lateral roots. Thick lateral roots can have more branches of roots, but fine lateral roots do not have any further branches. There is no secondary thickening on rice roots. The absorbance of water and nutrients is attributable to the fine lateral roots, while Thick Lateral Roots prove to be an aid in extending the root system as a whole for riveting water and essential nutrients into a larger quantity of soil (Gu et al., 2017).

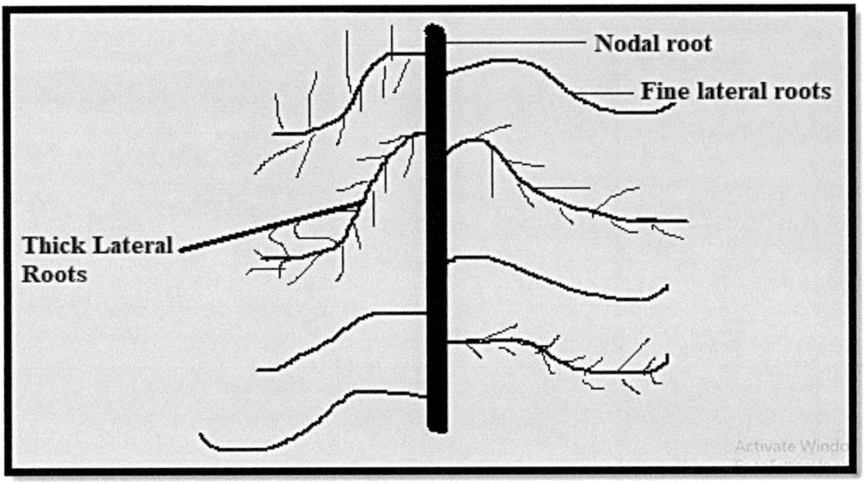

FIGURE 3.1 Morphological structure of rice root.

3.3 RHIZOSPHERE AND RICE

Rice is the uttermost and most important crop plant, whose root is composed of three distinct regions, endosphere, rhizoplane, and rhizosphere. Rice grows in flooded paddy fields and the aerenchyma tissue of the root releases oxygen leading to the rhizosphere which is relatively oxygen-rich and surrounded by anoxic bulk soil. The copiousness and diversity of microbes become different in the rice rhizosphere microbiome (Figure 3.2) than in the surrounding bulk soil due to rhizodeposition and sloughed-off root cells. In the rice rhizosphere, the bacterial population is most commonly present than other microbes. The diversity and shape of microbes are dependent on geographical location, type of soil, and genotype of rice as all these factors affect the pH of soil and root exudation pattern. Proteobacteria, Acidobacteria, Actinobacteria, and Chloroflexi phyla are the common bacterial population found in rice rhizosphere. Archaeal and fungi communities like Crenarchaeota, Thaumarcheotal, Euryarcheotic, and Chytridiomycota *Rhizophlyctis, Cladochytrium, and Aspergillus*, respectively are also commonly found in rice rhizosphere (Long-Jun Ding et al., 2019).

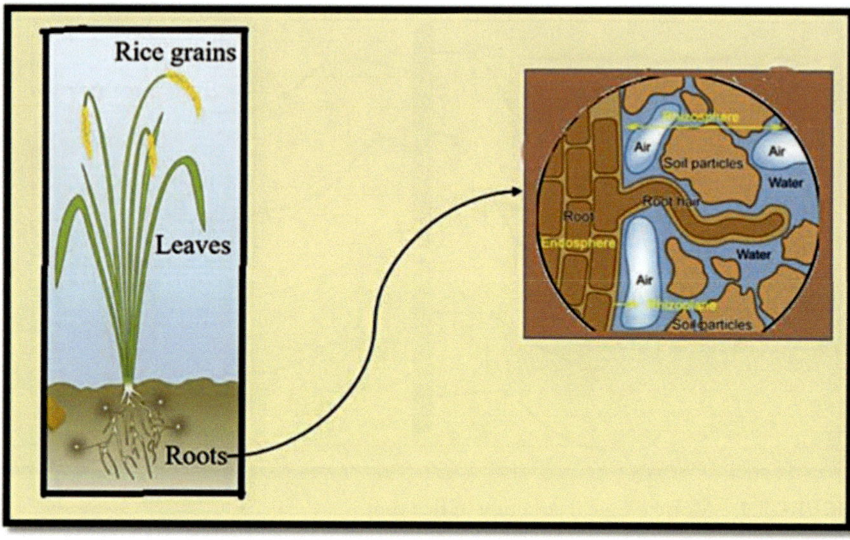

FIGURE 3.2 Rice rhizosphere microbiome.

One of the major factors that contribute to the diversity of rice rhizosphere is root exudates. Root exudates are regulated genetically and composed of amino acids, polysaccharides, and metabolites which attract microbes and promote their growth in the rice rhizosphere, making it unique and diverse. Root exudates act as a vital source of minerals and nutrients for microbes present in the rhizosphere and contribute to inducing their chemotactic response (Bacilio-Jiménez et al., 2003).

3.4 ARSENIC TRANSFORMATION

Arsenic (As) is a toxic substance that gets transported into rice grain from the soil. Rice is becoming a key provenance of arsenic in the world population through their diet as rice is consumed worldwide in one form or another. The arsenic biotransformation (Figure 3.3) is triggered by microbes present in the rhizosphere through the soil-rice system by these four main processes (Long-Jun Ding et al., 2019).

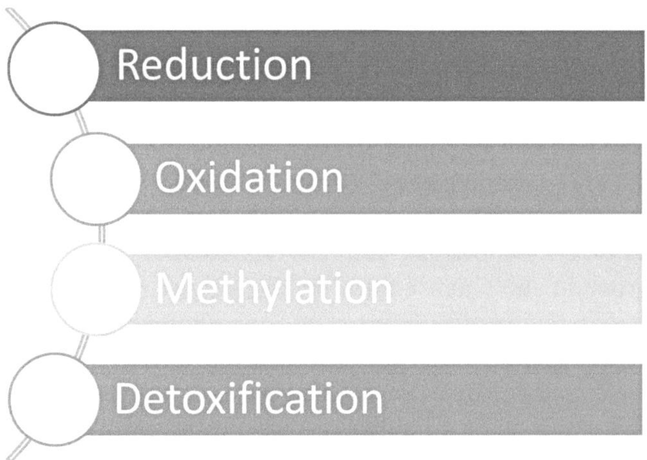

FIGURE 3.3 Process of arsenic biotransformation.

Arsenite (As III) – a form of arsenic is having more mobility in paddy soils in contrary to arsenate. The reduction of arsenate to arsenite is accomplished by iron and sulfate-reducing bacteria present in the rice rhizosphere microbiome. It has been found that arsenite can be oxidized by both heterotrophic and chemotrophic microbes and has been isolated from gold mines, soil with arsenic, sediment, geothermal environments, etc. A bacterial strain *Acidithiobacillus ferrooxidans*, has a tolerance for arsenic species, and aids in the mobilization of arsenic from pyrite (Rhine, 2008).

Microorganisms dwelling in the rice rhizosphere, produce DMA (Dimethylarsenic acid), which is required for As methylation as plants do not have the ability to produce it on their own and then being absorbed it through the roots (Jia et al., 2013). Arsenite methyltransferase (ArsM), methylating As III to organic Arsenic species, such as MMA (Monomethylarsonic acid), DMA, or volatile trimethylarsine (TMA), is generally disseminated in rice paddy soils. ArsM has been recognized from *Methanosarcina* sp. and also SM-1 from Cytophagaceae family (Hung et al., 2016) exhibited greater As methylation activity. In addition to, the direct influence on the biotransformation of arsenic, microorganisms residing in the rhizosphere also exhibited an indirect effect on the movement of arsenic in soil and its accumulation in rice grains via multiple processes related to cycles of C, N, and Fe and hence play essential roles in determining the fate of As in the soil-rice system.

3.5 TRANSPORTERS IN ARSENIC MOBILIZATION

There are various transporters that help in the mobilization of essential minerals and nutrients to different parts of the plants. Arsenate is less toxic than arsenite and is transported through the phosphate transporter family (PHTs). OsPHT1 (Kamiya et al., 2013) and OsPHT8 (Wu et al., 2011) are responsible for the uptake of arsenate in rice. A crucial transporter involved in the uptake of Arenate in rice plants is OsPT8. The conveyance of As(V) from the root to the shoot ensues by means of diverse PHT proteins. The two main regulators of phosphate transportation namely OsPHF1 (phosphate transporter traffic facilitator (i) and PHR2 (phosphate starvation response; and (ii) expressed their effect on arsenate uptake whereas arsenite and undissociated methylated species of arsenic are transmitted via aquaglyceroporins of diverse groups and more principally through NIP (Nodulin 26-like Intrinsic Protein) which belongs to family of aquaporins (AQP) channel (Awasthi et al., 2017). A NIP class transporter, OsNIP2;1 (Lsi1) is a prominent transporter of silicic acid, and plays a significant role in the uptake of arsenite (Ma et al., 2008). The transportation of arsenite is rice is facilitated by OsNIP1;1, OsNIP2;2 (OsLsi6), OsNIP3;1, OsNIP3;2, OsNIP3;3 in rice. OsHAC1;1 and OsHAC1;2 are localized in roots and play their role as arsenate reductases (ARs) (Shi et al., 2016). Recently Chen et al. (2017) explained an interesting fact regarding the role of OsNIP3;2 in the uptake mechanism of arsenite further it was found that the gene was predominantly confined in the lateral roots and stele region of the primary roots. Moreover, the mutation in the above-said gene is responsible for the reduced concentration of arsenite in roots but not in shoots. The process of As(III) efflux and xylem loading is mediated by OsLsi2 conveyed the participation of AQP from plasma membrane intrinsic proteins class including OsPIP2;4, OsPIP2;6, OsPIP2;7 (Mosa et al., 2012) and NRAMP1 (Tiwari et al., 2014) are involved in the uptake and transport of arsenite. Lsi is responsible for the uptake of organic forms of arsenic MMA and DMA. The role of transporters in the uptake of arsenic is described in Table 3.1.

3.5.1 LSI (LOW SILICON)

The first transporter for arsenite transportation in rice is Lsi, i.e., low silicone protein. The silicone and arsenite resemble each other in terms of their diameter and structure, so the transport system of Si is utilized by arsenite for its mobilization in rice plants. Silicone gets effluxed towards the stele region for xylem loading proficiently by Lsi1 and Lsi2, further, these two are also responsible for arsenite uptake in rice plants. A decrease of 57% is observed in AsIII quantity when a mutation occurs in Lsi1. Lsi1 and Lsi2 belong to the family of NIP, which is a subfamily of AQP in plants (Ma et al., 2008). A slow or reduced translocation of arsenite from root to shoot is found to be associated with the overexpression of OsNIP1;1 or OsNIP3;3 in rice plants (Singh et al., 2019).

3.5.2 ABCC (C-TYPE ATP-BINDING CASSETTE TRANSPORTERS FAMILY)

OsABCC1 belongs to the C-type ATP-binding cassette transporters family (ABCC) of the transporters class in *Oryza sativa*. These have a very significant role in the two main processes of arsenic in grains (rice) which are reduction and detoxification. OsABCC1 was confined in the vascular bundles, mainly in the phloem. The cold pack mutant of OsABCC1 leads to a significant decrease in the tolerance of arsenic in rice (Song et al., 2014). It was projected that OsABCC1 hinders the translocation of arsenic to the rice grains via the sequestration of arsenic in the vacuoles of the phloem companion calls at nodes in rice (Singh et al., 2019).

3.5.3 NRAMP (NATURAL RESISTANCE-ASSOCIATED MACROPHAGE PROTEINS)

NRAMPs are a family of integrated membrane proteins that facilitate divalent metal transport. Firstly, NRAMP was identified in mice (Vidal et al., 1993). As loading in the xylem is mediated by OsNRAMP1 and proposed that the gene can be a good candidate gene for evolving a new variety of rice with reduced arsenic content in grains via gene manipulation, and eventually reducing the jeopardy of As contamination in the food chain (Singh et al., 2019).

3.5.4 MATE (MULTIDRUG AND TOXIC COMPOUND EXTRUSION PROTEINS)

MATE stands for Multidrug and Toxic Compound Extrusion Proteins. This is a secondary cluster of active transporters facilitating the transport activity by the use of ATPases maintained by electrochemical gradient across the membrane (Kruda & Tsuchiya, 2009). The wracking of RNAi OsMATe2 via endosperm specific moderates arsenic accrual in grains and transgenic lines established a noteworthy decrease in OsMATE2 transcripts and arsenic content of grain in comparison to the wild variety of rice (Singh et al., 2019).

TABLE 3.1 Main Transporters and Their Role in Uptake of Arsenic

Sl. No.	Transporter Family	Gene/Protein	Role
1.	Phosphate transporter	OsPht1, OsPht8	Uptake of As V in rice (Kamiya et al., 2013; Wu et al., 2011)
2.	Aquaporins	Lsi 1	Uptake and transport of As(III).
		OsNIP2;1	Arsenite efflux and xylem loading (Singh et al., 2019)
		OsNIP1;1, OsNIP2;2	
		OsLsi6, OsNIP3;1 OsNIP3;2, OsNIP3;3	
		Lsi2	
3.	NRAMP	OsNRAMP1	As III uptake and transport (Singh et al., 2019)
4.	ABCC	OsABCC1	Hinders translocation of arsenic to grains by vacuole sequestration (Singh et al., 2019)
5.	MATE	OsMATE2	Moderates arsenic accumulation in rice grain (Singh et al., 2019)

3.6 UPTAKE AND TRANSPORT KINETIC OF ARSENIC SPECIES IN RICE

3.6.1 INORGANIC ARSENIC (ASI) SPECIES

Arsenite and Arsenate are the two main inorganic arsenic (Asi) (Figure 3.4). There are two mechanisms that come into the role in the uptake of inorganic As species in rice plants. The high-affinity phosphate transporter is responsible for the transmittance of arsenate As(V) from roots to aerial parts of the plant. The gene family of phosphate transporters consists of 13 OsPT gene transporters varying from OsPT1 to OsPT13 (Paszkowski et al.,

2003). Among all the 13 genes the two important genes which contribute to the uptake of arsenic are OsPT1 and OsPT8. OsPT8 is responsible for arsenate absorption into the roots of rice plants, and its over-expression results in the accretion of arsenic in roots, leading to profound toxic effects (Wang et al., 2016). OsPT1 is responsible for the uptake of arsenate from roots to shoots.

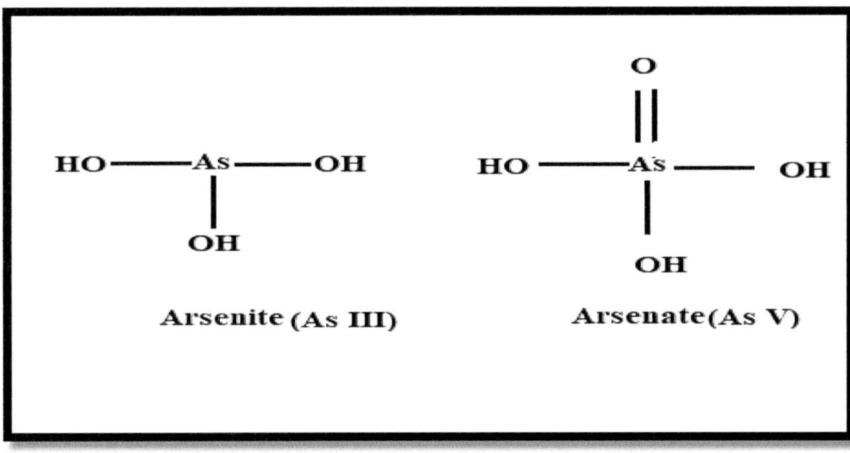

FIGURE 3.4 Structure of inorganic form of arsenic.

Another mechanism is the uptake of arsenic species As(III) through the AQP channel. In this mechanism, silica with its influx major transporter (OsNIP2;1) and efflux transporters of Si play a significant role. In the xylem, arsenite is transported through the medium of Lsi1, a nodulin 26 alike intrinsic protein (OsNIP2;1), a primary influx transporter for Si (Li et al., 2009; Ma et al., 2006) and protein Lsi2, an efflux transporter of Si (Gonzalez, 2014). The Lsi protein channel performs the bidirectional function in such a way that some portion of arsenite that is taken up by the roots is immediately released back into the rhizosphere. AsIII is the reduced form of As(V) and is requisitioned into root vacuoles followed by its translocation to the shoots and it is circulated to different parts of the plant (Anindita Mitra et al., 2017).

3.6.2 *ORGANIC SPECIES OF ARSENIC*

The prominent form of organic arsenic is MMA and DMA (Figure 3.5). The inorganic species of As are transformed into organic species of As by the microbes present in the rhizosphere producing methylated species of arsenic such as DMA and MMA (Meharg et al., 2009). The methylated form of arsenic is less toxic than the inorganic form, although the uptake of arsenic in its inorganic form is more rapid by the plant roots than the methylated species but the translocation rate of methylated arsenic in aerial parts is more than inorganic As. The uptake of MMA and DMA is facilitated via the NIP. The improved translocation of methylated As is attributable to its reduced complex formation with the ligands GSH/PC. In rice grain arsenite are the predominant species followed by DMA, arsenate, and MMA whereas in rice straw As(III) is found in abundance followed by arsenite and DMA (Anindita Mitra et al., 2017).

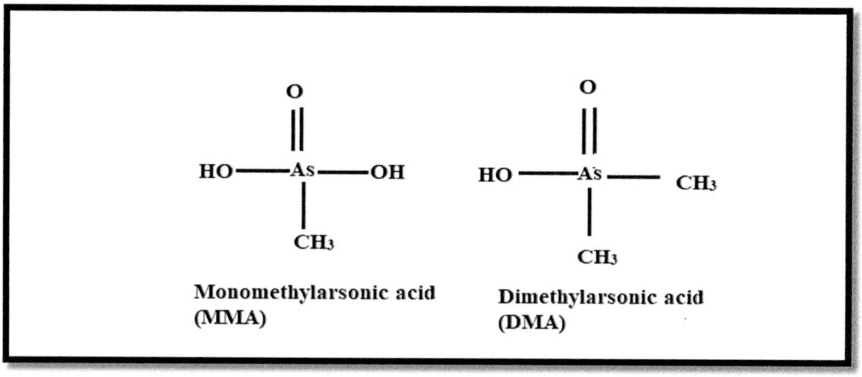

FIGURE 3.5 Structure of organic forms of arsenic.

3.7 ARSENIC DETOXIFICATION IN RICE

The peptide thiol such as glutathione and phytochelatins have sulfhydryl bonds (-SH). Arsenite is having high affinity to sulfuryl bonds. The detoxification of arsenite is mediated by the complex formation of As with GSH or PCs (Pal & Rai, 2010). The process of synthesis of glutathione involves two steps catalyzed by the rate-limiting γ-EC (γ-glutamylcysteine) synthetase to synthesize γ-EC, trailed by glutathione synthetase to combine Gly and γ-EC (Hell & Bergmann, 1990; Dhankher et al., 2002). The transpeptidation of gamma-glutamyl cysteinyl dipeptides from GSH catalyzed by means of PC synthase activity is the process involved in the synthesis of phytochelatins (Pal & Rai, 2010; Shri et al., 2014). The over-expression of phytochelatin synthase increased the tolerance for As in plants and also its accumulation in plant parts., OsCLT1, plays a significant part in the homeostasis of GSH, possibly via facilitating the transfer of γ-EC and GSH to the cytoplasm from plastids. Furthermore, under arsenic treatment, mutants of OsCLT1 parades reduced contents of PC2 contrary to the wild one, ensuing reduced accrual of arsenic in the root part but greater or parallel arsenic accretion in the shoots (Yang et al., 2016). OsABCC1 in rice is crucial for arsenic detoxification and As III – PCS vacuolar sequestration. So, the cold pack of $OsABCC_1$ leads to increased sensitivity of arsenic (Chen Yanshan et al., 2017).

3.8 ARSENIC SEQUESTRATION IN RICE

The uptake of AsIII and AsV by rice roots is performed by AQP and transporters of phosphate (Figure 3.6). In rice, inorganic arsenic AsIII with phytochelatin is sequestered in a vacuole as a footstep of detoxification of As and affect arsenic distribution in different tissue of the plant. $OsABCC_1$ is present in roots and expressed specifically in exodermis and pericycle arbitrates vacuolar AsIII–PC confiscation, thus dropping arsenic percentage in rice grains (Song et al., 2014). In the shoots, a cold pack of OsABCC1 in shoot lowers the percentage of As in node I, but upsurges allocation of arsenic to the grain and leaf of the plant, resulting in a higher concentration of arsenic in brown rice in contrary to the wild type. In the roots, Vacuolar sequestration can serve as a blockade to restrain the metallic translocation in the shoot of the plant (Ueno et al., 2010). The critical step in the allocation of As in vacuoles is the formation of a complex of As III by thiols concurrently articulating the ABC group of transporters and PC synthase the ABC transporters and PC synthase, the rate-limiting step in the biogenesis of phytochelatins, may increase sequestration of arsenic (Chen Yanshan et al., 2017).

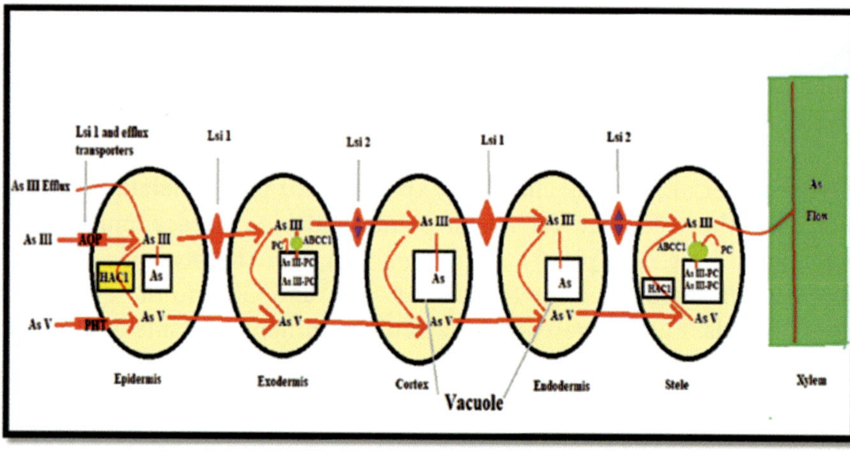

FIGURE 3.6 Uptake of arsenite and arsenate and its sequestration in vacuoles (root).

Description of Figure 3.6 – the uptake of AsIII and AsV by rice roots is performed by AQP and transporters of phosphate. AR HAC1 reduces arsenate to arsenite. The influx and efflux transporters of arsenite, i.e., Lsi1 and Lsi2, respectively play a vital part in the transportation of arsenic to the xylem (root). Arsenite can form a complex with phytochelatins (PC) and ABCC1 mediates the transport of the AsIII-PC complex that confines to the membranes of vacuoles and further transport arsenic into the vacuoles (root) for its sequestration. Lsi1 with other transporters intercede arsenite efflux.

3.9 ARSENIC METHYLATION AND VOLATILIZATION

Arsenic methylation as a part of arsenic detoxification is extensive in bacteria, fungi, and archaea as they are proficient in the methylation of inorganic species, as well as transmuting arsenite into arsines which are volatile in nature. The conversion of the Asi to monomethylarsonous acid (MMAs(III)), followed by dimethylarsinous acid (DMAs(III)) dimethylarsinous acid (DMAs(III)), and eventually to the volatile TMAs is known as arsenic methylation (Figure 3.7). A gene encoding As III S-adenosylmethionine methyltransferase (ArsM) from the bacterium *Rhodopseudomonas palustris* (Qin et al., 2006) speeds up the formation of intermediates (methylated in nature) namely DMA^{V} and TMAO from inorganic As III and TMAsIII gas as a final product.

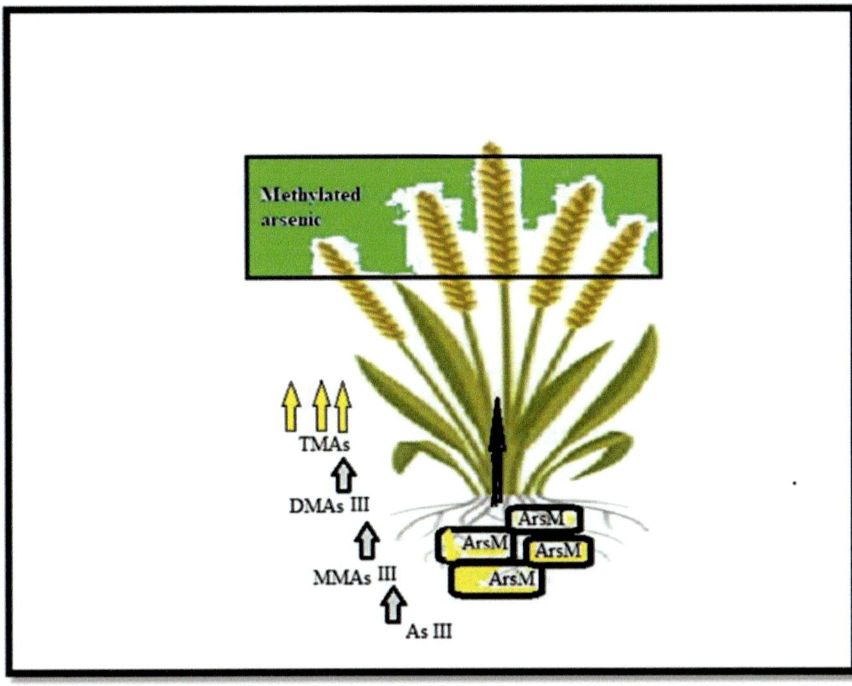

FIGURE 3.7 Methylation process of arsenic in rice.

In several soil environments and plants, different methylated As varieties are found, out of which MMAs, DMAs, and TMAs are likely to be vaporized from the soil into the environment. However, much information is not known about the microorganisms that catalyze the crucial pathway of the As cycles in the soil environment. Methylated As species is usually present in rice grain prominently in the form of (DMAs(V)), moderately as (MMAs(V)), (TMAs(V)O), and seldomly as (TETRAs(V)). Rhizosphere plays a significant role in this as Methylated As species are quite likely to be aroused via rhizosphere, through the activities of methylating microorganisms, and not from invocating methylation in rice (Figure 3.8). The rice rhizosphere varies from the surrounding soil in terms of different properties such as physiochemical, composition, and activities of microbes. This is because of the release of oxygen and transudation of mucilage, polysaccharides, amino acids, and organic acids from the root part, highlighting the significance of rhizospheric approaches in the transformation of arsenic and its uptake into rice plants (Chen et al., 2017).

Though microbial methylation of arsenite for the first time has been observed in 1935. It was found that the process of methylation in arsenic is speed by arsenite methyltransferase (ArsM) including the transfer of methyl group from S-adenosyl-L-methionine (SAM) to arsenite. The operons' resistance to arsenic, although varies in diverse groups of bacteria and archaea, is present in a majority of microorganisms, and confers intransigence to arsenite to arsenate. Volatilization of arsenic into arsines may bequeath to the global cycle of Arsenic. The process of methylation and volatilization of arsenic in rice may get catalyzed by the regulation of the ArsM gene, resulting in lesser accretion of toxic arsenic in rice grain (Yan et al., 2013).

Biomethylation of Arsenic has been measured as an efficient detoxification technique of microbes ensuring the formation of trivalent arsenic species which are methylated in nature. MMA and DMA are mediators in the biosynthetic course of pentavalent methylated arsenic (MMA^V and DMA^V), further, they are the most noxious forms of arsenic. Trivalent form of arsenic differs from their pentavalent complements in terms of structure, moreover, they are more reactive and less carcinogenic. It has been reported that the trivalent form of arsenite and MMA act via similar mechanisms to produce toxicity, further a steadier and more stable binary complex is formed by MMA^{III} in the active site of the enzyme in contrast to arsenite (José et al., 2016).

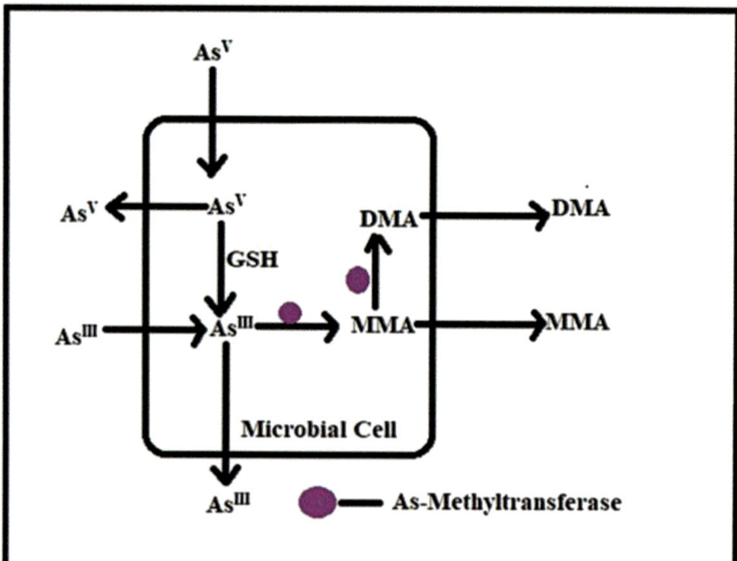

FIGURE 3.8 Ways of methylation of arsenic by microorganisms and its excretion.

3.10 ARSENIC MOBILITY

Arsenic mobility in rice or any other plant is affected by its species. Organic As species have higher root-to-shoot mobility than an inorganic form of arsenic. The arsenic is concentrated more in the root ensuring a low shoot/root ratio ranging between 0.1 and 0.3 (Zhao et al., 2009). Arsenite and arsenate have similar transporters for their with P and Si with, Silicic acid having higher mobility in contrast to arsenic with the ratio of the concentration of xylem sap to that in the external atmosphere being around 20 and 0.3–0.6 for Si and As, respectively. The reason for this difference in mobility is due to the complex formation of arsenic with phytochelatins, whereas no such complex formation takes place in the case of phosphate and silicic acid. The higher arsenic accrual in rice as compared to other crop plants is attributable to the localization of transporters of arsenic (influx and efflux) in exodermis and endodermis (Su et al., 2009), leading to efficient transportation to xylem sap. Arsenite in the form of AsIII/AsV is the most prominent species found in xylem sap, with a pattern in which initially the concentration of As V remains higher for the first 2 hours, and then gradually the concentration of As III starts increasing. The uptake of As V in the xylem is increased by the fundamental expression of transporters from the phosphate family (OsPht1;8) (Wu et al., 2011). In the case of the organic group, only pentavalent species MMA(V) and DMA(V) are found to be present in both phloem and xylem sap with higher concentrations in phloem sap. As was found to be stockpiled in phloem in the top node along with the internode (Awasthi et al., 2017).

3.11 ACCUMULATION OF ARSENIC INTO RICE GRAIN

The transportation of nutrients takes place by means of ovular vascular trace to endosperm via chalaza and nucellar projection (Krishnan & Dayanandan, 2003). The transportation of arsenic follows the same route as that of essential nutrients. The nourishing cut rice panicles with arsenite, or DMA transported the Asi or DMA to the emerging plant caryopsis. The DMA transportation is more efficient as compared to As III. When phloem is removed in the process of stem girdling the concentration of As III in rice grain decreased by 90%, whereas a reduction of 55% is observed in DMA concentration of rice grain. This demonstrates that transportation of As III prominently occurs by means of phloem whereas in the case of DMA, both xylem and phloem are likewise vital for its transportation into the grain (Carey et al., 2010). Arsenite is found in the ovular vascular trace region whereas DMA is present in endosperm, their localization is one of the main reasons for the difference in their mobility and transportation to the rice grain. $OsPTR_7$ a putative peptide transporter in rice has an important role in long-distance transportation from root to shoot and in translocation of DMA to rice grain (Awasthi et al., 2017).

The profile of arsenic in rice plants in terms of its accumulation, speciation, distribution, and concentration get affected by many factors such as genotype and environmental factors. It was determined by Ahmed et al. (2011). That the variation in As concentration in rice grain is affected most prominently by environment, genotype, and genotype-environment. Syu et al. (2015) explained that the arsenic concentration and species in six genotypes of rice in Taiwan were either similar or greater in genotypes Indica as compared to the Japonica genotype. One of the most important steps in the direction of enhancing the production of safe cultivars of rice in arsenic-smitten fields/areas is the identification of genotypic variation.

3.12 IRON PLAQUE FORMATION AND RADIAL OXYGEN LOSS AFFECTING AVAILABILITY OF ARSENIC AT ROOT LEVEL

The roots of the rice plant are having extensive aerenchyma which allows infiltration of oxygen from the shoot required for the respiration of roots, further in submerged paddy soils the anaerobic condition arises, and therefore, a part of oxygen is released within the rhizosphere by the aerenchyma. This whole process is termed radial oxygen loss and varies according to the genotype of the plant, water logging condition, and availability of oxygen in soil (Colmer et al., 2006). Iron plaque develops on the surface of roots due to the oxidation of ferrous iron (Fe^{2+}) to ferric iron (Fe^{3+}) leading to the precipitation of iron oxides/hydroxides. The iron plaque thus formed is orange in color and is a major place for the storage of arsenic because of the formation of iron oxides/hydroxides which are the efficient sorbents for arsenic. This decreases the concentration of arsenic in rice roots. X-ray absorption near-edge structure (XANES) can be practiced discerning the arsenic species in iron plaque, mostly As V is found in iron plaque. The impact of iron plaque on As V is controversial as some studies demonstrated that it acts as a barrier to As V (Liu et al., 2004), and other studies suggested that it acts as a source of As V (Liu et al., 2006). The contribution of Fe plaque in prompting concentration of arsenic may differ as per the deposition of iron plaque on the surface of the root that undeniably fluctuates from the tip of the roots to the oldest roots and from primary to lateral ones root tips to old roots and from primary to lateral roots. In the rice cultivars, which release more oxygen from the roots, the iron plaque formation is increased resulting in lesser uptake of arsenic, and as more oxygen is released As III gets converted to arsenate which is readily get adsorbed on Fe plaque (Figure 3.9). The translocation of As is reduced to shoot due to the high binding capacity of iron oxides and hydroxides to As V. Recently, the microbial composition has been found to be a crucial factor for determining As species existing on the iron plaque. Complementary variances were present in the composition of microbes and multiplicity between plaque formation and rhizospheric soil. Arsenite-oxidizing bacteria existing on an iron plaque in roots (*Acidovorax and Hydrogenophaga*) were found to be intricated in As transformation and henceforward affect arsenic absorption in rice tissues (Hu et al., 2015).

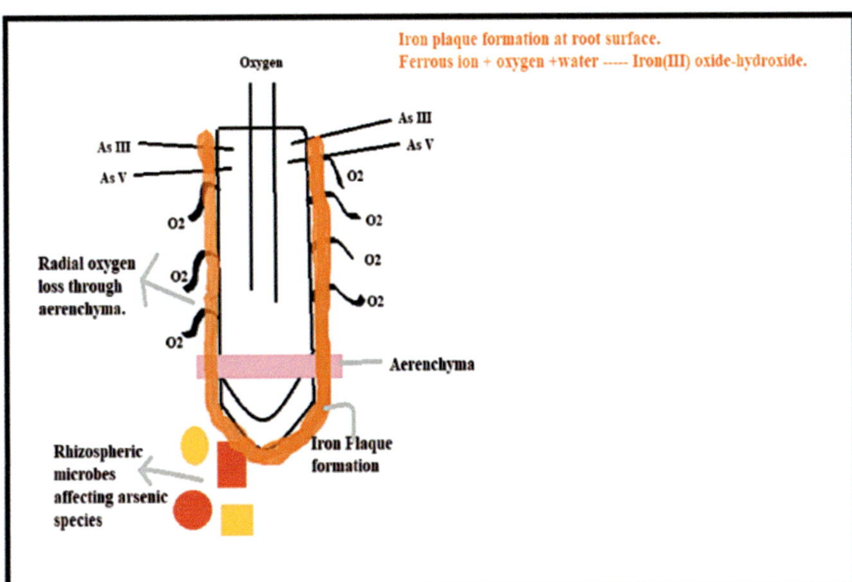

FIGURE 3.9 ROL and formation of iron plaque on the surface of roots and their influence on the availability of arsenic for its uptake.

3.13 INFLUENCE OF SULFUR ON ARSENIC TRANSPORTATION

Sulfur (S) hinders the transportation, absorption, adsorption, and translocation of arsenic in rice. The greater the supply of Sulfur, the lower the concentration of arsenic in rice. The high supply of S regulated the expression of many genes that are involved in the transportation of arsenic. Sulfur facilitates the down-regulation of P transporters OsPT23 and aquaporin gene: OsTIP4;2 whereas upregulation of ABC transporters (OsABCG5, OsABC17_2, and OsABC6) and PC synthase genes (OsPCS1, OsPCS3, and OsPCS13) suggesting that greater supply of S permits plants to produce more phytochelatins and sequestration of As to vacuoles proficiently by formation of a complex with phytochelatins via upregulation of phytochelatins and ABD genes (Zhang et al., 2016). The application of S also appears to lower the uptake of arsenite and arsenate at the surface of roots via the downregulation of P transporters and the AQP channel, respectively (Awasthi et al., 2017).

3.14 ROLE OF PHOSPHOROUS AND SILICIC ACID IN ARSENIC TRANSLOCATION

Phosphorous and arsenate are analogous to each other in terms of structure. Phosphorous competes with arsenate along with root adsorption of oxides and iron hydroxides, therefore, the concentration of As decreases as the concentration of phosphorous increases. Studies conducted on different types of soil with different concentrations of iron and phosphorous suggest a positive relation amidst the concentration of iron and As in grains of rice but a negative association in the concentration of phosphorous and arsenic (Jiang et al., 2014). The arsenate toxicity increases as inhibition of oxidative phosphorylation along with the replacement of arsenate in place of phosphorous in ATP, interfere with the synthesis of ATP (Rahman et al., 2014b). Various elements and their importance in arsenic physiology are described in Table 3.2.

In the case of silicic acid, similarities were observed between silicic acid and arsenious acid, allowing the arsenious acid to penetrate the root by means of silicone transporters. The reason behind the increased concentration of antioxidant enzymes along with their isoenzymes and decreased arsenic phytotoxicity is the presence of silicone. It also increases the concentrations of cysteine, GSH, and NPSH, and lowers As-induced lipid peroxidation in rice. An effective strategy to reduce arsenic accumulation is silicone fertilization (José et al., 2016).

TABLE 3.2 Elements and Their Importance in Arsenic Physiology

Sl. No.	Element	Importance
1.	Iron (Fe)	Iron plaque formation acts as a storage site for arsenic.
		Iron act as a strong absorbent of arsenic, further availability of arsenic gets increased due to the release of adsorbed arsenic in the reducing environment by reductive dissolutions of iron oxyhydroxides. Bacteria associated with the reduction process of Fe are reported to be linked with arsenic mobilization.
2.	Phosphorous (P)	The transport of arsenic in different forms is facilitated by Phosphate transporters, further As also interferes with the metabolism pathway of phosphate.
3.	Sulfur (S)	S aids in the detoxification of arsenic and also maintains the concentration of arsenic in the root, further limiting its translocation to tillers.
4.	Silicone (Si)	The use of Silicone fertilizers in arsenic-contaminated soil is an effective measure to reduce arsenic accrual in rice.

3.15 FACTORS AFFECTING ARSENIC INTAKE IN RICE PLANT

The interaction that takes place within the soil between microbes, chemicals, and soil is affected by a number of factors such as physical, biological, and chemical properties of soil, the microclimate of soil, microbes present in the soil, and many more. The factors that affect the mobilization of arsenic in rice plants are discussed in subsections (Figure 3.10 and Table 3.3).

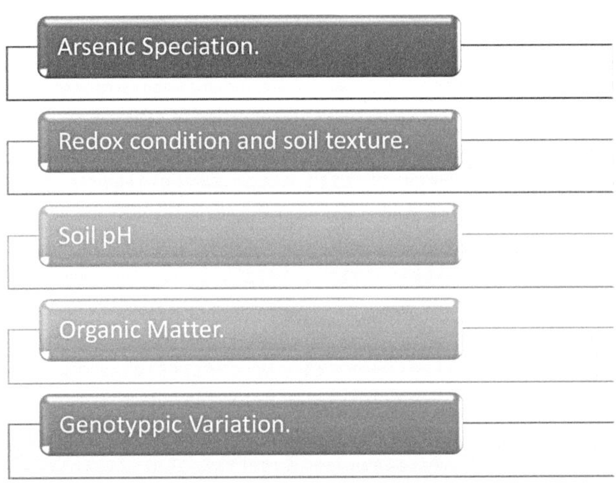

FIGURE 3.10 Factors affecting arsenic intake in rice plant.

3.15.1 ARSENIC SPECIATION

As is found in both the forms of soil – organic and inorganic. As(V) and As(III) are the most common inorganic form while MMA, i.e., monomethyl arsenic acid, and DMA, i.e., dimethylarsinic acid are the common organic form of arsenic (Figure 3.11). Arsenite is the most toxic form of arsenic followed by arsenate, MMA, and DMA (Baig et al., 2010). In submerged paddy fields, where the surrounding conditions are anaerobic, arsenite is the most prominently found. It was stated by Marin et al. that the bioavailability of arsenic species in rice plants follows the orders As III > MMA >AsV > DMA (Marin et al., 1992). The rate of uptake of inorganic species is more velociously than the uptake rate of organic species.

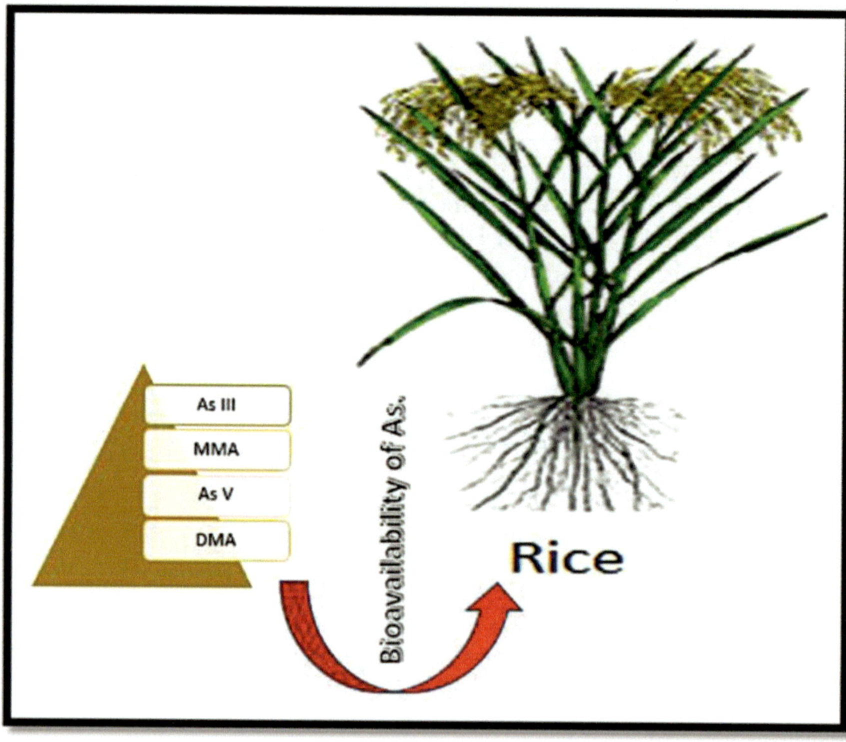

FIGURE 3.11 Order of availability of arsenic in the biosphere to rice plant.

3.15.2 EFFECT OF REDOX CONDITION AND SOIL TEXTURE

Redox chemistry plays a very significant part in mobilization and As accrual into rice plants (Williams et al., 2007). In aerobic conditions, i.e., Oxidized soil conditions, arsenic prevails as As V, and further gets adsorbed on Fe-oxyhydroxide stages which act as a barrier in the availability of arsenic to rice (Lauren et al., 2005). In reduced or anaerobic condition arsenic prevail as arsenite and become available to rice plant for uptake by means of the dissolution of iron oxides (Takahashiand et al., 2004) also due to the decrease of arsenate to arsenite by microbial process (Islan et al., 2004). Nearly some microbial communal inhabitants of rice rhizosphere can upsurge the availability of As in the biosphere by the dissolution of ferric ion by exuding siderophores to the interface of rot-plaque resulting in 160 mg kg^{-1} arsenic concentration in the root of rice plants (Bhattacharya et al., 2007).

Arsenic solubility and bioavailability are also get affected by the texture of soil (Adriano et al., 2001). Finer texture and larger surface area are the characteristics of silt and clayey soil allowing Fe oxides, resulting in higher arsenic scavenging potential and lower levels of arsenic toxicity in soil. The situation is reversed in the case of sandy and loamy soil as the arsenic phytotoxicity is five times lower in both soils (Quazi et al., 2011).

3.15.3 EFFECT OF SOIL PH

Soil pH determines the arsenic speciation and leaching, hence affecting its mobility and bioavailability to rice plants. Fe-oxyhydroxide an arsenic-binding species become more soluble at a $pH<5$ (Signes-Pastor et al., 2007), increasing arsenic uptake in rice. At $pH = 8.5$ the ions containing negative charge at the surface increase, for example, hydroxyl ions, enabling desorption of As from iron oxides resulting in the As mobilization in the root part, which, in turn, augments As accretion in plants (Ahmed et al., 2011).

3.15.4 EFFECT OF ORGANIC MATTER

The soil organic matter regulates the availability of arsenic to the plant. It was reported by Pikaray et al. (2005) that organic matter has an enhanced affinity for the adsorption of arsenic, leading to the formation of the organo-arsenic complex. In soils having high organic matter due to the formation of organo-arsenic complex, the solubility of arsenic gets reduced affecting its uptake in the plant. Various research studies reported the same finding that the higher the organic matter content in the soil, the lesser the arsenic concentration in rice plants. An interesting fact about arsenic mobility in soil with high organic matter is that by enhancing microbial activity and lowering the redox potential of soil, arsenic solubilization can be increased (Turpeinen et al., 1999), an environment fortunate for the reductive suspension of Fe-oxyhydroxides allied to organic matter (Selim et al., 2010).

3.15.5 EFFECT OF GENOTYPIC VARIATION

Different varieties of rice is having different arsenic accumulation rates in rice grains. The highest accumulation of arsenic concentration, i.e., 1.77 mg kg^{-1} is found in the BR11 variety (Meharg et al., 2003) and the lowermost arsenic concentration was reported in the cultivar of JX-17. The different rice varieties such as IR 50, White Minikit, and Red Minikit were having arsenic concentration ranges within 0.24–0.31 mg kg^{-1} and considered efficient accumulators of arsenic in comparison to Nayanmani, Jaya, Ratna, Ganga-Kaveri, and Lal Sanna whose arsenic accumulation range lies within 0.14–0.20 mg kg^{-1}; with a maximal accumulation of 0.31 mg kg^{-1} in White Minikit, and the minimal of 0.14 mg kg^{-1} in Jaya (Bhattacharya et al., 2010). 'TN1' and 'ZYQ8' cultivar of rice has the largest arsenic concentration in vegetative parts and root whereas the cultivar 'JX-17' was about 50% arsenic accumulation in roots and vegetative parts as that to cultivar 'ZYQ8' (Zhang et al., 2008). The difference in concentration of arsenic in different arsenic varieties is due to many factors such as environmental, genetic, percentage of arsenic in irrigation water and soil, root anatomy, porosity, Fe plaque formation, interaction within the rhizosphere, and arsenic tolerance gene (Anindita et al., 2017).

TABLE 3.3 Factors and Their Effect on Arsenic Intake in Rice Plants

Sl. No.	Factor	Effect on Arsenic Uptake
1	Arsenic speciation	The rate of uptake of inorganic species is faster than the organic species' uptake rate.
2.	Redox condition and soil texture	In oxidized soil conditions, arsenic prevails as As(V) and further gets adsorbed on Fe-oxyhydroxide stages which act as a barrier in the availability of arsenic to rice.
		In reduced or anaerobic condition arsenic prevail as arsenite and become available to rice plant for uptake.
		Sandy and loamy soil favors the phytotoxicity of arsenic as compared to silt and clayey soil.
3	Soil pH	At $pH<5$ arsenic uptake in rice plants increases.
		At $pH = 8.5$ the ions containing negative charge at the surface increases, such as hydroxyl ions, enabling desorption of arsenic from Fe-oxides resulting in the mobilization of arsenic in the root area, which, in turn, augments arsenic accretion in the plant.
4.	Organic matter	Due to the formation of an organo-arsenic complex, the solubility of arsenic gets reduced affecting its uptake in the plant.
5.	Genotypic variation	Different varieties of rice have different arsenic accumulation rates.

3.16 CONCLUSION

A number of issues are emerging regarding the rise of As quantity in rice and attributes affecting the uptake, transportation, translocation, and arsenic accrual in rice. Phosphate transporters, AQP, and Si transporters are the gateway for arsenic facilitating its entry in either inorganic or organic form into plants. Methylated arsenic is less toxic than inorganic species of arsenic, further in organic species, arsenite is more toxic than arsenate. Accumulation of arsenic in rice basically depends on its bioavailability and is directed by several attributes like soil types, physicochemical parameters, presence of other elements, and mineral composition such as sulfur and silicon in soil, soil-rhizosphere-plant system rhizospheric microorganisms and their activities; organic matters, iron-plaque formation, radial oxygen loss, etc. Amendments in agronomic activities and bioremediation strategies may be a feasible method to lower arsenic accretion in rice. Some methodological approaches like crop irrigation by rainwater harvesting, use of arsenic-resistant microorganisms in bioremediation, application of natural arsenic chelators, enhancing the production of genetically modified rice plants, and the aerobic farming of paddy crops are efficacious to alleviate arsenic contamination from rice (Anindita et al., 2017).

KEYWORDS

- **arsenic**
- **arsenic speciation**
- **iron plaque**
- **pH**
- **rhizosphere**
- **rice**
- **silicone transporters**

REFERENCES

Adriano, D. C., (2001). *Trace Elements in the Terrestrial Environments: Biogeochemistry Bioavailability, and Risks of Metals* (pp. 47–71). Springer: New York, USA.

Ahmed, Z. U., Panaullah, G. M., Gauch, H., McCouch, S. R., Tyagi, W., Kabir, M. S., et al., (2011). Genotype and environment effects on rice (*Oryza sativa* L.) grain arsenic concentration in Bangladesh. *Plant Soil, 338*, 367–382. doi: 10.1007/s11104-010-0551-7.

Awasthi, S., Chauhan, R., Srivastava, S., & Tripathi, R. D., (2017). The journey of arsenic from soil to grain in rice. *Frontiers in Plant Science, 8*, 1007. doi=10.3389/fpls.2017.01007. https://www.frontiersin.org/article/10.3389/fpls.2017.01007 (accessed on 28 March 2023).

Bacilio-Jiménez, M., Aguilar-Flores, S., Ventura-Zapata, E., et al., (2003). Chemical characterization of root exudates from rice (*Oryza sativa*) and their effects on the chemotactic response of endophytic bacteria. *Plant and Soil, 249*, 271–277. https://doi.org/10.1023/A:1022888900465.

Baig, J. A., Kazi, T. G., Shah, A. Q., Kandhro, G. A., Afridi, H. I., Khan, S., & Kolachi, N. F., (2010). Biosorption studies on powder of stem of *Acacia nilotica*: Removal of arsenic from surface water. *J. Hazard. Mater., 178*, 941–948.

Bhattacharya, P., Mukherjee, A. B., Bundschuh, J., Zevenhoven, R., & Loeppert, R., (2007). *Arsenic in Soil and Groundwater Environment: Biogeochemical Interactions, Health Effects and Remediation* (Vol. 9). Elsevier: Amsterdam, The Netherlands.

Bhattacharya, P., Samal, A. C., Majumdar, J., & Santra, S. C., (2010). Uptake of arsenic in rice plant varieties cultivated with arsenic rich groundwater. *Environ. Asia, 3*, 34–37.

Carey, A. M., Scheckel, K. G., Lombi, E., Newville, M., Choi, Y., Norton, G. J., et al., (2010). Grain unloading of arsenic species in rice. *Plant Physiol., 152*, 309–319. doi: 10.1104/pp.109.146126.

Chen, Y., Yong-He, H., Cao, Y., Yong-Guan, Z., Rathinasabapathi, B., & Ma Lena, Q., (2017). Arsenic transport in rice and biological solutions to reduce arsenic risk from rice. *Frontiers in Plant Science, 8*, 268. doi: 10.3389/fpls.2017.00268. https://www.frontiersin.org/articles/10.3389/fpls.2017.00268/full (accessed on 28 March 2023).

Colmer, T. D., Cos, M. C. H., & Voesenek, L. A. C. J., (2006). Root aeration in rice (*Oryza sativa*): Evaluation of oxygen, carbon dioxide, and ethylene as possible regulators of root acclimatizations. *New Phytol., 170*, 767–777. doi: 10.1111/j.1469-8137.2006.01725.

Dhankher, O. P., Li, Y., Rosen, B. P., Shi, J., Salt, D., Senecoff, J. F., et al., (2002). Engineering tolerance and hyperaccumulation of arsenic in plants by combining arsenate reductase and γ-glutamylcysteine synthetase expression. *Nat. Biotechnol., 20*, 1140–1145. doi: 10.1038/nbt747.

Gonzalez, P. S., Talano, M. A., Oller, A. L. W., Ibanez, S. G., Medina, M. I., & Agostini, E., (2014). Update on mechanisms involved in arsenic and chromium accumulation, translocation and homeostasis in plants. In: Gupta, D. K., & Chatterjee, S., (eds.), *Heavy Metal Remediation: Transport and Accumulation in Plants* (pp. 45–72). Nova Science Publishers: Hauppauge, NY, USA.

Gu, D., Zhen, F., Hannaway, D. B., Zhu, Y., Liu, L., Cao, W., & Tang, L., (2017). Quantitative classification of rice (*Oryza sativa* L.) root length and diameter using image analysis. *PloS One, 12*(1), e0169968. https://doi.org/10.1371/journal.pone.0169968.

Hell, R., & Bergmann, L., (1990). γ-Glutamylcysteine synthetase in higher plants: Catalytic properties and subcellular localization. *Planta, 180*, 603–612. doi: 10.1007/BF02411460.

Hu, M., Li, F., Liu, C., & Wu, W., (2015). The diversity and abundance of As(III)-oxidizers on root iron plaque is critical for arsenic bioavailability to rice. *Sci. Rep., 5*, 13611. doi: 10.1038/srep13611.

Huang, K., Chen, C., Zhang, J., et al., (2016). Efficient arsenic methylation and volatilization mediated by a novel bacterium from an arsenic-contaminated paddy soil. *Environ. Sci. Technol., 50,* 6389–6396.

Islam, F. S., Gault, A. G., Boothman, C., Polya, D. A., Charnock, J. M., Chatterjee, D., & Lloyd, J. R., (2004). Role of metal reducing bacteria in arsenic release from Bengal delta sediment. *Nature, 430*, 68–71.

Jia, Y., Huang, H., Zhong, M., et al., (2013). Microbial arsenic methylation in soil and rice rhizosphere. *Environ. Sci. Technol., 47*, 3141–3148.

Kamiya, T., Islam, R., Duan, G., Uraguchi, S., & Fujiwara, T., (2013). Phosphate deficiency signaling pathway is a target of arsenate and phosphate transporter OsPT1 is involved in as accumulation in shoots of rice. *Soil Sci. Plant Nutr., 59*, 580–590. doi: 10.1080/00380768.2013.804390.

Krishnan, S., & Dayanandan, P., (2003). Structural and histochemical studies on grain-filling in the caryopsis of rice (*Oryza sativa* L.). *J. Biosci., 28*, 455–469. doi: 10.1007/BF02705120.

Lauren, J. G., & Duxbury, J. M., (2005). Management strategies to reduce arsenic uptake by rice. In: *Proceedings of the International Symposium on Behavior of Arsenic in Aquifers, Soils and Plants: Implications for Management*. Dhaka, Bangladesh; Centro Internacional de Mejoramiento de Maíz y Trigo and the U.S. Geological Survey: Reston, VA, USA.

Li, R. Y., Ago, Y., Liu, W. J., Mitani, N., Feldmann, J., McGrath, S. P., Ma, J. F., & Zhao, F. J., (2009). The rice aquaporin Lsi1 mediates uptake of methylated arsenic species. *Plant Physiol., 150*, 2071–2080.

Liu, W. J., Zhu, Y. G., Hu, Y., Williams, P. N., Gault, A. G., Meharg, A. A., et al., (2006). Arsenic sequestration in iron plaque, its accumulation and speciation in mature rice plants (*Oryza sativa* L.). *Environ. Sci. Technol., 40*, 5730–5736. doi: 10.1021/es060800.

Liu, W. J., Zhu, Y. G., Smith, F. A., & Smith, S. E., (2004). Do phosphorus nutrition and iron plaque alter arsenate (As) uptake by rice seedlings in hydroponic culture? *New Phytol., 162*, 481–488. doi: 10.1111/j.1469-8137.2004.01035.

Long-Jun, D., Hui-Ling, C., San-An, N., Xi-En, L., Gui-Lan, D., & Yong-Guan, Z., (2019). Microbiomes inhabiting rice roots and rhizosphere. *FEMS Microbiology Ecology, 95*(5), fiz040. https://doi.org/10.1093/femsec/fiz040.

Ma, J. F., & Yamaji, N., (2006). Silicon uptake and accumulation in higher plants. *Trends Plant Sci., 11,* 392–397.

Marin, A. R., Masscheleyn, P. H., & Patrick, W. H., (1992). The influence of chemical form and concentration of arsenic on rice growth and tissue arsenic concentration. *Plant Soil, 139*, 175–183.

McNear, Jr. D. H., (2013). The rhizosphere – roots, soil and everything in between. *Nature Education Knowledge, 4*(3), 1.

Meharg, A. A., & Rahman, M. D. M., (2003). Arsenic contamination of Bangladesh paddy field soils: Implications for rice contribution to arsenic consumption. *Environ. Sci. Technol., 37*, 229–234.

Meharg, A. A., Williams, P. N., Adomako, E., Lawgali, Y. Y., Deacon, C., Villada, A., Cambell, R. C. J., Sun, G., Zhu, Y. G., Feldmann, J., et al., (2009). Geographical variation in total and inorganic arsenic content of polished (white) rice. *Environ. Sci. Technol., 43*, 1612–1617.

Mitra, A., Chatterjee, S., Moogouei, R., & Gupta, D. K., (2017). Arsenic accumulation in rice and probable mitigation approaches: A review. *Agronomy, 7*, 67.

Mosa, K. A., Kumar, K., Chhikara, S., Mcdermott, J., Liu, Z., Musante, C., et al., (2012). Members of rice plasma membrane intrinsic proteins subfamily are involved in arsenite permeability and tolerance in plants. *Trans. Res., 21*, 1265–1277. doi: 10.1007/s11248-012-9600-8.

Pal, R., & Rai, J. P., (2010). Phytochelatins: Peptides involved in heavy metal detoxification. *Appl. Biochem. Biotechnol., 160*, 945–963. doi: 10.1007/s12010-009-8565-4.

Paszkowski, U., Kroken, S., Roux, C., & Briggs, S. P., (2002). Rice phosphate transporters include an evolutionarily divergent gene specifically activated in arbuscular mycorrhizal symbiosis. *Proc. Natl. Acad. Sci. USA, 99*, 13324–13329.

Pikaray, S., Banerjee, S., & Mukherji, S., (2005). Sorption of arsenic onto Vindhyan shales: Role of pyrite and organic carbon. *Curr. Sci., 88*, 1580–1585.

Qin, J., Rosen, B. P., Zhang, Y., Wang, G., Franke, S., & Rensing, C., (2006). Arsenic detoxification and evolution of trimethylarsine gas by a microbial arsenite

S-adenosylmethionine methyltransferase. *Proc. Natl. Acad. Sci. U.S.A., 103*, 2075–2080. doi: 10.1073/pnas.0506836103.

Quazi, S., Datta, R., & Sarkar, D., (2011). Effect of soil types and forms of arsenical pesticide on rice growth and development. *Int. J. Environ. Sci. Technol., 8*, 450–460.

Rahaman, S., Sinha, A. C., & Mukhopadhyay, D., (2011). Effect of water regimes and organic matters on transport of arsenic in summer rice (*Oryza sativa* L.). *J. Environ. Sci., 23*, 633–639.

Rhine, E. D., Onesios, K. M., Serfes, M. E., Reinfelder, J. R., & Young, L. Y., (2008). Arsenic transformation and mobilization from minerals by the arsenite oxidizing strain WAO. *Environmental Science & Technology, 42*(5), 1423–1429. https://doi.org/10.1021/es071859k.

Selim, R. A. H. M., Jena, J. S., Yang, H. J., Lee, M. K., Woodall, B., Liu, C. C., Lee, J. F., & Luo, S. D., (2010). Occurrence of arsenic in core sediments and groundwater in the Chapai-nawabgang District, Northwestern Bangladesh. *Water Res., 44*, 2021–2037.

Shi, S., Wang, T., Chen, Z., Tang, Z., Wu, Z., Salt, D. E., et al., (2016). OsHAC1; 1 and OsHAC1; 2 function as arsenate reductases and regulate arsenic accumulation. *Plant Physiol., 172*, 1708–1719.

Shri, M., Dave, R., Diwedi, S., Shukla, D., Kesari, R., Tripathi, R. D., et al., (2014). Heterologous expression of *Ceratophyllum demersum* phytochelatin synthase, CdPCS1, in rice leads to lower arsenic accumulation in grain. *Sci. Rep., 4*, 5784. doi: 10.1038/srep05784.

Signes-Pastor, A., Burló, F., Mitra, K., & Carbonell-Barrachina, A. A., (2007). Arsenic biogeochemistry as affected by phosphorus fertilizer addition, redox potential and pH in a West Bengal (India) soil. *Geoderma, 137*, 504–510.

Song, W. Y., Yamaki, T., Yamaji, N., Ko, D., Jung, K. H., Fujii-Kashino, M., et al., (2014). A rice ABC transporter, OsABCC1, reduces arsenic accumulation in the grain. *Proc. Natl. Acad. Sci. U.S.A., 111*, 15699–15704. doi: 10.1073/pnas.1414968111.

Su, Y. H., McGrath, S. P., & Zhao, F. J., (2010). Rice is more efficient in arsenite uptake and translocation than wheat and barley. *Plant Soil, 328*, 27–34. doi: 10.1007/s11104-009-0074-2.

Takahashi, Y., Minamikawa, R., Hattori, K. H., Kurishima, K., Kihou, N., & Yuita, K., (2004). Arsenic behavior in paddy fields during the cycle of flooded and non-flooded periods. *Environ. Sci. Technol., 38*, 1038–1044.

Tiwari, M., Sharma, D., Dwivedi, S., Singh, M., Tripathi, R. D., & Trivedi, P. K., (2014). Expression in *Arabidopsis* and cellular localization reveal involvement of rice NRAMP, OsNRAMP1, in arsenic transport and tolerance. *Plant Cell Environ., 37*, 140–152. doi: 10.1111/pce.12138.

Turpeinen, R., Pantsar-Kallio, M., Haggblom, M., & Kairesalo, T., (1999). Influence of microbes on the mobilization, toxicity and biomethylation of arsenic in soil. *Sci. Total Environ., 236*, 173–180.

Ueno, D., Yamaji, N., Kono, I., Huang, C. F., Ando, T., Yano, M., et al., (2010). Gene limiting cadmium accumulation in rice. *Proc. Natl. Acad. Sci. U.S.A., 107*, 16500–16505. doi: 10.1073/pnas.1005396107.

Wang, P., Zhang, W., Mao, C., Xu, G., & Zhao, F. J., (2016). The role of OsPT8 in arsenate uptake and varietal difference in arsenate tolerance in rice. *J. Exp. Bot., 67*, 6051–6059.

Williams, P. N., Villada, A., Deacon, C., Raab, A., Figuerola, J., Green, A. J., Feldmann, J., & Meharg, A. A., (2007). Greatly enhanced arsenic shoot assimilation in rice leads to elevated grain levels compared to wheat and barley. *Environ. Sci. Technol., 41*, 6854–6859.

Wu, Z., Ren, H., McGrath, S. P., Wu, P., & Zhao, F. J., (2011). Investigating the contribution of the phosphate transport pathway to arsenic accumulation in rice. *Plant Physiol., 157*, 498–508. doi: 10.1104/pp.111.178921.

Yan, J., Hai, H., Min, Z., Feng-Hua, W., Li-Mei, Z., & Yong-Guan, Z., (2013). *Environmental Science & Technology, 47*(7), 3141–3148. doi: 10.1021/es303649v.

Yang, J., Gao, M. X., Hu, H., Ding, X. M., Lin, H. W., Wang, L., et al., (2016). OsCLT1, a CRT-like transporter 1, is required for glutathione homeostasis and arsenic tolerance in rice. *New Phytol., 211*, 658–670. doi: 10.1111/nph.13908.

Zhang, J., & Duan, G. L., (2008). Genotypic difference in arsenic and cadmium accumulation by rice seedlings grown in hydroponics. *J. Plant Nutr., 31*, 2168–2182.

Zhang, J., Zhao, C. Y., Liu, J., Song, R., Du, Y. X., Li, J. Z., et al., (2016). Influence of sulfur on transcription of genes involved in arsenic accumulation in rice grains. *Plant Mol. Biol. Rep., 34*, 556–565. doi: 10.1007/s11105-015-0937.

Zhao, F. J., Ma, J. F., Meharg, A. A., & McGrath, S. P., (2009). Arsenic uptake and metabolism in plants. *New Phytol., 181*, 777–794. doi: 10.1111/j.1469-8137.2008.02716.

CHAPTER 4

Long-Distance Translocation Mechanism of Arsenic from Soil to Rice Grain

MANJU SHRI,[1] SONALI DUBEY,[2] SANJAY DWIVEDI,[3,4] PRADYUMNA KUMAR SINGH,[3,5] and RUDRA DEO TRIPATHI[3,4]

[1]*School of Applied Sciences and Technology, Gujarat Technological University, Ahmedabad, Gujarat, India*

[2]*Institute of Biotechnology, Amity University, Noida, Uttar Pradesh, India*

[3]*Council of Scientific and Industrial Research–National Botanical Research Institute (CSIR-NBRI), Lucknow, Uttar Pradesh, India*

[4]*Academy of Scientific and Innovative Research (AcSIR), Ghaziabad, Uttar Pradesh, India*

[5]*School of Earth and Environmental Sciences, Babasaheb Bhimrao Ambedkar University, Lucknow, Uttar Pradesh, India*

ABSTRACT

Arsenic is a non-essential toxic element and hazardous for the growth and development of both plants and animals. Arsenic is efficiently taken by rice plants due to the presence of Lsi and phosphate transporters and accumulated in various parts such as root, shoot, leaf, and grain. The mechanisms of As mobilization from soil to root and transportation to different parts of rice plants depend upon various factors, including soil pH, availability

Arsenic in Rice: A Global Menace. Garima Awasthi, Sudhakar Shrivastava, and Mahipal Singh Sankhla, (Eds.)

of nutrients and As species in soil, and transporters of As translocation. Reduction of AsV into AsIII by OsHAC1;1 and OsHAC1;2 and subsequent binding of AsIII with thiolic ligands for vacuolar sequestration in root is the vital mechanism to reduce the As toxicity and translocation from root to shoot. Further, transporters, for instance, OsLsi2, OsLsi6, OsPht1;8, OsNRAMP1, OsPT2, and OsPTR7 involved translocation of various species of As from root to shoot may be potential targets to minimize the As mobilization towards the shoot. Arsenic accumulation in rice grain is mainly contributed by phloem loading and unloading. Inositol and OsPTR7 transporters are taking part in the accumulation of As species in rice grain. However, the detailed mechanism of As mobilization shoots to grain and factors regulating the As accumulation in grain is needed to be elucidated. In this chapter, we highlighted the mechanisms which play an important role in As mobilization from soil to grain and have the potential to minimize the As accumulation in rice grain for safer food production.

4.1 INTRODUCTION

Arsenic (As) is a non-threshold, class 1 human carcinogen (Int. Agency Res. Cancer, 2004), and its bioavailability to crop plants is a serious health threat to the human population. Rice is a dominant source of inorganic As, for the population based on the rice diet, who are not exposed to highly As-contaminated drinking water. For the population exposed to high As in drinking water, such as in the Bengal Delta region, rice accounts for approximately 50% of inorganic As intake (Mondal & Polya, 2008). Through arsenic (As) contaminated irrigation water and soil, As enters crop and fodder plants, resulting in the contamination of human food (Meharg, 2004; Williams et al., 2005; Tripathi et al., 2007). Also illustrating this point is the World Health Organization's (WHO) provisional maximum tolerable daily intake (MTDI) limit of As of 2 μg As Kg^{-1} body weight per day (WHO, 1993). If rice grain with an As level of 0.1 μg As g^{-1}, which is typically found in the literature (William et al., 2005), is consumed at a rate representative of a rice-based subsistence diet (0.42 kg dry weight per day; Baffes & Gautam, 2001), then ingestion of 0.7 μg Kg^{-1} body weight occurs for a 60 Kg human, which contributes 35% of the MTDI. Zavala & Duxbury (2008) derived a global normal range of 0.08–0.2 μg g^{-1} As in rice based on samples from rice-growing countries around the world.

Levels far in excess in rice grain have been repeatedly reported for West Bengal, India, and Bangladesh.

Arsenic exists predominantly as arsenate (AsV) and arsenite (AsIII) in nature. Arsenate is an analog of phosphate and is taken up by plants via the high-affinity phosphate transport systems (Meharg & Macnair, 1992; Abedin et al., 2002; Zhao et al., 2009). Transport of AsIII is known to occur through aquaglyceroporins as a neutral molecule in rice (Abedin et al., 2002; Isayenkov & Maathuis, 2008; Ma et al., 2008). Accumulation of As in rice depends on many factors such as climate, soil type, and rice variety. Onken & Hossner (1995) grew rice in two soil types simulated with up to 45 μg AsIII or AsV g^{-1} for 60 days. The As concentration of rice plants correlated with the mean soil solution AsV concentration in the clay soil and to the mean soil solution AsIII for the silt loam. The rate of As uptake by plants increased as the rate of plant growth increased. Abedin et al. (2002) studied the uptake kinetics of different species of As in rice plants. Organic As species, monomethyl arsenic acid (MMA), and dimethylarsinic acid (DMA) showed much lower rates of uptake than AsV and AsIII. Phosphate application did not affect AsIII uptake and toxicity but reduced AsV uptake and toxicity. DMA and MMA also showed restricted translocation with 0–5% of total As in rice straw being DMA. When irrigated with AsV-contaminated water, rice plants have been found to accumulate As to levels of 1.8 μg g^{-1} dw in seeds and up to 92 μg g^{-1} As in rice straw. It has been demonstrated that Lsi1, a silicon transporter, transports AsIII and also mediates the transport of MMA and DMA in rice (Li et al., 2009b). Extensive investigations indicate that As bioavailability in rice is highly dependent on As speciation, which in turn can vary depending on rice cultivar and As in irrigation water (Juhasz et al., 2006). In the present scenario, the decontamination of the As from soil and irrigation water is not a feasible approach using available physical and chemical remediation methods (Mondal et al., 2006). Therefore, to mitigate the problem without compromising the nutritional quality of rice, investigation of rhizospheric As chemistry is with the identification of As tolerant cultivars having low grain As is needed.

The bioavailability of As in soil is usually affected by the physico-chemical properties of Fe. There are few reports on the effects of interactions between AsV, phosphate, and iron hydroxides/oxides on As behavior in soil-plant systems. Iron plaque is commonly formed on the surfaces of roots of aquatic plants, including *Oryza sativa,* due to the oxidation of ferrous to ferric iron and the precipitation of iron oxide on the root surface, promoted by the release of oxygen and oxidants into the rhizosphere. The importance of iron plaque in As attenuation in the rhizosphere has been demonstrated in rice plants (Liu et al., 2004; Tripathi et al., 2014). The study on phosphorus-As interactions in the soil-rice system (Liu et al., 2004) revealed that P nutrition of rice plants might to some extent control the formation of iron plaque, and this phenomenon, in turn, may influence AsV uptake by rice plants. Arsenic concentrations in iron plaque were significantly higher in –P plants than in +P plants. Concentrations of As in shoots were significantly lower in –P plants than in +P plants. This indicates that iron plaque might sequestrate As, and consequently reduce the translocation of As from roots to shoots. Chen et al. (2005) found that iron plaque had contrasting effects on the uptake of AsIII and AsV. While uptake of AsIII was enhanced by the presence of iron plaque, that of AsV was dramatically inhibited. Thus the role of rice rhizospheric chemistry is important to know the level of arsenic toxicity in rice grain and its effects on the nutritional quality of rice. The human population is increasing day by day and the land utilized for crop production is decreasing either as a result of pollution or as a result of industrialization. There is a need to increase crop production to fulfill the increased amount of food supply. In this study, the detailed mechanism of As mobilization from soil to rice grain will be focused on. Further, the key checkpoints which have the potential to minimize the accumulation of As in rice grain will also be discussed. The knowledge which is gathered from various research areas and groups had made a better understanding of As bioavailability its uptake, transport, detoxification, loading to xylem, phloem, and accumulation in rice grain, which can be utilized for mitigating the toxic effects of As accumulation in rice grain.

4.2 SOURCES OF ARSENIC CONTAMINATION AND ITS BIOAVAILABILITY TO RICE

Ground-water contamination by human-generated or natural sources with numerous contaminants such as As has now developed into a serious environmental concern worldwide. Large numbers of populations in different countries are wide-open to elevated amounts of As consumption by using As-contaminated groundwater (Ravenscroft et al., 2009). Presently, As is assessed to impact around 150 million people globally through drinking water (Stroud et al., 2011). Transmission of As in the food chain will subsequently persist as an enduring hazard to social and environmental systems (Tuli et al., 2010). The knowledge of groundwater As contamination along with its associated risks to health and As a detoxification mechanism is required as Arsenic can enter into the human body through As polluted water (Shaji et al., 2021).

The existence of As appears in over 200 minerals, and the decomposition of As from naturally existing As-containing minerals leads to the As elevation in the underground water (Bissen & Frimmel, 2003). Excessive As concentration has been found in groundwater where As occurs in specific rock-forming minerals, for example, pyrite and arsenopyrite (Ghosh & Singh, 2009). Arsenic concentration in sediments was found to be ranged from 3 to 10 mg/kg depending on the minerals of different natural and textural components (Chakraborty et al., 2015). The iron oxide, hydrous metal oxide, or pyrites containing land exhibit very high levels of As in sediments when compared with other oxides (Shrivastava et al., 2015). Increased concentration of As is found in reducing sediments which rises progressively with the proliferating depth of the sediment (Charlet et al., 2007). Iron, as well as Al oxides containing sediments, are also responsible for contaminating the groundwater. Aluminum and iron oxide reductive dissolution together with the metal-reducing bacteria directly effectuates the movement of As (Jang et al., 2016). Combustion of mineral fuels and coal, mining, application of herbicides, arsenical fungicides, and insecticides in agricultural lands, and wood preservers are major anthropogenic sources that play an important role in contaminating ground-water with As (Figure 4.1) (Nriagu et al., 2007). Coal flaming causes volatilization of As_4O_6 which leads to the release of As in the atmosphere, and is further eventually transported into water reserves (Bissen & Frimmel, 2003).

Due to disease prevention and growth promotion effects, P-arsanilic acid (AsA) has been extensively utilized as a regular supplement in animal feed for broiler chickens and pigs (Mangalgiri et al., 2015). In pig and chicken antibiotics, the concentration of AsA has been identified at around 45 and 30 mg per kg, respectively (Straw et al., 2002). The presence of organic acid as AsA in 25.4% of 146 animal feeds was reported by Yao et al. (2013) with an average concentration of 21.2 mg kg^{-1}. Even though USA and China are not utilizing As in animal feed anymore, however, As in soils can be found due to the long-term application of poultry litter containing As. Additionally, AsA can be degraded into more toxic compounds when untreated waste used as organic fertilizer in agricultural lands as well as animal manure are not managed responsibly (Wang et al., 2014). Inorganic arsenic (Asi) is released into the soil from the environment and flows into the water bodies which can be taken up by crops and eventually humans can be exposed to the As (Chung et al., 2014). As a result, long-term exposure to As may pose a great risk to human health through food chains due to the accumulation of this element within plants growing in contaminated areas. Though, the threat exhibited by different As forms in rice rely on the bioavailability of metal.

Rice (*Oryza sativa*) is the principal cereal grain grown worldwide and works as a staple dietary source over and above half of the population around the world (Fukagawa & Ziska, 2019). In Bangladesh, almost 80% of energy consumption is supplied by rice locally (Shew et al., 2019). Rice is more effective in the accumulation of As than other crops because the anaerobic surroundings in immersed soils result in the movement of AsIII that is absorbed by rice roots with the help of the silicic acid pathway (Mitra et al., 2017). Thus, rice is a main source of Asi for the population using rice as a staple food (Islam et al., 2017). Arsenic in rice grain is mainly existing in the form of methylated As, specifically DMA (dimethyl arsenic acid) (Arao et al., 2011). Pentavalent methylated arsenicals are less toxic than Asi. The relative percentage of Asi and DMA varies in rice grains according to rice genotypes (Norton, 2009) and the conditions used to grow rice (He et al., 2021). The ecological condition of soil affects the availability of As species and their accumulation in rice plants. The current evidence suggested that As is methylated by microbes using arsenite methyl transferase. The methylated species are taken up by rice roots through aquaporins (AQP). The As accumulation of rice grain, its negative impact on yield, and subsequent transmission to the food chain are of main concern (Murugaiyan et al., 2021). The As-containing water used for irrigation comprises mostly soluble Asi. Khan et al. (2010) speculated that the use of As-containing water for irrigation purposes contaminated the paddy soils of the Bengal delta region and possess a greater bioavailability of As in soils in comparison to soil either holding geogenic As or polluted by other sources such as mining procedure. Moreover, the supplementation of As in these soils is the latest procedure permitting an adequate period for As to age in soil, particularly underneath anaerobic environments in the rice field, which are favorable to As mobilization (Xu et al., 2008). As mobilization and speciation in soil pore water, rice growth, and As uptake was compared by Khan et al. (2010) in 14 soils from the greenhouse of China, Bangladesh, and the United Kingdom to investigate As speciation in rice grain based on soils and growing surroundings.

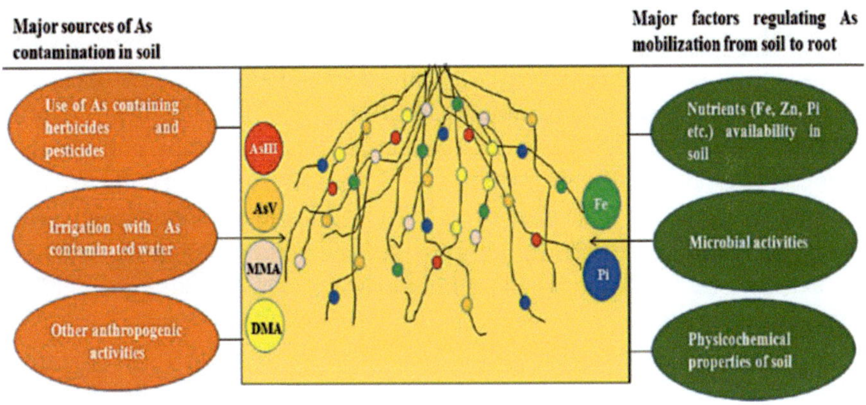

FIGURE 4.1 Factors affecting arsenic (As) accumulation in soil and mobilization from soil to rice root.

4.3 FACTORS INFLUENCING THE MOVEMENT OF AS FROM SOIL TO PLANTS

In soil, the availability of As depends on its source and form. Characteristics and components of soil can regulate the solubility and toxicity of As (Hussain et al., 2021). Different parameters influence the bioavailability of As in soils, such as pH, ionic strength of the soil solution, organic carbon content, iron oxides, and cation exchange capacity (Romero-Freire et al., 2014). The degree of As solubility in water is more efficiently associated with As content in the plant compared with the As level in the soil. In rice crops, soil conditions, for example, iron hydr/oxides concentration, level of As, plant accessible P (phosphorus), availability of silicon to plant, and quality of soil affected As accumulation. Nutrients and pollutant availability can deviate through the rhizosphere as soil settings can fluctuate from one place to another (Figure 4.1).

4.3.1 EFFECT OF PH

Soil pH regulates the As solubility, mobility, and bioavailability. Arsenic has a huge affinity for oxides/hydroxides in soil. Soil pH based on the organic matter, soil quality, and minerals composition can influence the As uptake from particular soil constituents to plants (Azam et al., 2016). The low pH of the soil can also be responsible for low As extraction in comparison to other metals such as zinc and cadmium (Száková et al., 2009). Red mud is the fine waste residue obtained after extracting Al from bauxite using the Bayer Process. It has been reported that the AsV mobilization in red mud was greatly reliant on pH as well as As discharge to the solution was highly enhanced after supplementing the red mud suspension with phosphate (Lockwood et al., 2014). Al, Mn, and hydrous oxides of metal strongly take up both AsV and AsIII in acid soils whereas anionic As types are absorbed by calcium oxides in the alkaline situation (Shrivastava et al., 2015). In addition, As speciation and solubility were observed to be regulated by pH which can enhance the phytoavailability of As in contaminated regions (Gersztyn et al., 2013).

In aerobic soil environments under pH 8.5, less AsV mobilization is reported due to its efficient adsorption on different mineral elements comprising Fe and Al(hydr)oxides and aluminosilicates. In contrast around $pH > 6$, the sorption ability of ferric (hydr)oxide has been observed higher for AsIII than for AsV (Goldberg, 2002). Furthermore, the soil-to-solution separation of As is strongly controlled by the pH and dissolved volumes of the Fe-bearing phase in rice paddy fields (Yamaguchi et al., 2011).

4.3.2 INFLUENCE OF MICROORGANISMS

It has been shown that As tolerance can be enhanced by AMF (arbuscular mycorrhizal fungi) in plants (Sharma et al., 2017). Li et al. (2016) reported the reduced transport of inorganic/organic As to the rice grains in the presence of AMF. Under aerobic soils, inoculation of rice with AMF ameliorates the soil nutrients like N (nitrogen) and P (phosphorous) with growth, altering the As transport to the root (Yeasmin et al., 2007). In addition, considerable effects of AMF on root biomass, grain production, AsV, AsIII, DMA, total As concentrations, propositions of inorganic as well as organic As content in rice grains and soil As concentrations of rhizosphere have been identified (Li et al., 2016). In rice grains of all varieties, AMF significantly reduced the ratio of inorganic/organic As concentrations by the conversion of Asi into less toxic DMA and supported the As detoxification in rice.

Microbial oxidation of AsIII is considered sustainable remediation due to less toxicity and mobility of AsV than AsIII (Wang et al., 2015). Diverse classes of AsIII-oxidizing bacteria and archaea such as green sulfur bacteria, filamentous green non-sulfur bacteria, α-,β-,γ-Proteobacteria, *Thermus*, Crenarchaeota, and Euryarchaeota have been reported (Li et al., 2013). Moreover, the oxidation of AsIII is used as an energy reservoir by a few AsIII oxidizing bacteria (Ghosh et al., 2014). AsIII is used as the electron donor by *Rhizobium* sp. NT-26 (Chemolithoautotrophic AsIII-oxidizing bacteria) along with O_2 as the electron acceptor as well as CO_2/bicarbonate for the carbon resource to support the growth of bacteria (Santini et al., 2000). Furthermore, *Hydrogenophaga* sp. N14 and *Agrobacterium tumefaciens* GW4 (heterotrophic AsIII-oxidizing bacteria) also utilize AsIII oxidation as a source of energy (Shi et al., 2020). In paddy soils, identification, and functional analysis of microbes involved in As methylation is considerable for determining the DMA origination in the grains of rice. Bacterial strain GSRB54 has a strong ability to methylate As and was isolated from the rice roots growing in As-polluted paddy soil under anaerobic conditions (Kuramata et al., 2015). Heterologous expression of GSRB54 arsM in *E. coli* enhanced the methylation of AsIII by the conversion of AsIII into DMA and trimethylarsine oxide. In addition, DMA was detected in rice shoots grown in a liquid medium containing AsIII along with GSRB54 inoculum. It was speculated that the oxidative zone is colonized by the GSRB54 strain around the roots of rice and is associated with the accumulation of DMA as Streptomyces are generally aerobic bacteria (Kuramata et al., 2015).

4.3.3 NUTRIENTS IN THE SOIL

The oxides of Fe and Mn form under a toxic environment. The affinity among Fe, Mn oxides, and As enhances the growing redox potential of sediments (Martin, 2009). pH and redox potential of sediments play an important role in the stability of iron as well as iron hydr/oxides. In an acidic atmosphere, Fe commonly exists in soluble forms, whereas precipitation of Fe minerals is promoted by neutral pH in oxic soils (Du Laing et al., 2009). Xu et al. (2017) studied that two important features regulating As mobility are As linked with Fe oxy/hydroxides and the Mn oxides concentration in the flooded paddy soil environment. Supplementation of Mn oxide in paddy soils reduces As mobilization and As uptake in rice plants, indicating a promising alleviation approach to inhibit the As transportation from paddy soil to the food chain. Furthermore, Fe executes a critical role in the distribution of As distribution in rice land (Liu et al., 2006). The As mobilization happens through the surface hydroxyl group substitution with the As ions during its sorption to iron oxides and via amorphous ferric arsenates or insoluble secondary oxidation minerals synthesis (Kumpiene et al., 2008). In addition, surface complexation reaction on oxides or hydroxides of Al, manganese (Mn), and mostly iron (Fe) regulates the activity of As in soil (Deschamps et al., 2003).

Hossain et al. (2008) reported the spatial distribution of As content of the soil, irrigation water as well as rice crop in a tube-well govern region and their affinity with Fe and Mn. Over the stretch from the low tube-well point, the authors identified that approximately 68–136 μg/L As concentrations of water in the 110 m lengthy irrigation canal markedly decreased. The such deteriorating movement was also stated with Fe strengths (Hossain et al., 2008). Reviewing soil As the As content exposed a declining trend with the gap from the tube well pump additionally. As accumulation in grains of rice is associated with As in soil for example both extractable and total As (Hossain et al., 2008).

4.4 TRANSLOCATION AS FROM SOIL TO ROOT TO ABOVE-GROUND PARTS

Plants differ very much in their efficiency for transporting As within the different parts. The efficiency of As translocation from root to above-ground parts affects the tolerance capacity and accumulation of As in plants. Hyper accumulator plants exhibit higher translocation factors for metal(loid)s and can accumulate a higher amount of metal(loid)s without showing any abnormality in its growth. They also have higher efficiency of As translocation from root to above-ground parts and accumulating more As in leaf in comparison to root. In contrast to As hyperaccumulator, the crop plants have low efficiency of As translocation from root to shoot.

However, the crop plants accumulate the As in above-ground edible parts. Consequently, decoding the mechanism and transporters involved in the long-distance transport of As from root to above-ground part might be important for As remediation (phytoremediation) as well as for reducing As accumulation in edible parts of crop plants, thus enhancing food safety.

4.4.1 ARSENIC TRANSPORT FROM SOIL TO ROOT

In paddy soil conditions, arsenite (AsIII) and arsenate (AsV) are the dominant As species which is followed by methylated As species. AsV predominates in the aerobic condition of paddy soil. However, AsIII is the predominant species in the submerged condition (paddy rice cultivation). Different transporters are being utilized for the transport of As species by plants. AsV shows structural as well as chemical similarity with inorganic phosphate and thus is taken up by plants via phosphate transporters. Various phosphate transporters have been identified and characterized that transport AsV from soil to the root in various plants.

In *Arabidopsis thaliana* AtPHT1;1, AtPHT1;4, AtPHT1;5, AtPHT1;7, AtPHT1;8, and AtPHT1;9 have important roles in transporting AsV from soil to the roots. AtPht 1;1 and AtPht 1;4 are involved in As transport from soil to roots under increased or decreased concentrations of Pi (Shin et al., 2004). Furthermore, AtPht1;5 has the potential to uptake more As from As source and transport it to sink. AtPht1;7, AtPht1;8 and AtPht1;9 appreciably enhanced As and P uptake in the presence or absence of Pi in soil solution which further leads to augmentation in As toxicity inside the plants (LeBlanc et al., 2013). In rice 13 Pi transporters (OsPT) have been identified that can transport AsV. Along with Pi transporters, transcription factors also regulated the transport of AsV inside the root. Wang et al. (2016) demonstrated the involvement of OsPT8 in AsV transport to the root cells and in reducing root elongation as a result of As toxicity. Knockout lines of OsPHT1;4 showed a reduction in the transport of AsV and further accumulation of As in rice grain (Cao et al., 2017). The transcription factor such as OsPHR2, and WRKY regulated AsV transport in rice plants. In rice, OsWRKY28 and OsPHF1 (phosphate transporter traffic facilitator 1) play a key role in AsV uptake (Chen et al., 2011; Wang et al., 2018).

AsIII has a structural resemblance with silicic acid at neutral pH. In rice plants, silicic acid is very much required for the strengthening of rice leaf blades. The transporters of silicic acid are thus highly upregulated. AsIII and undissociated methylated As species are transported with the transporters that transport the neutral molecules such as water, and ammonia. NIPs (nodulin 26-like intrinsic proteins) are a class of aquaporins that are mainly responsible for AsIII transport. In rice, OsNIP2; 1 (Lsi1), a transporter for silicic acid (Si) has a major role in AsIII uptake (Ma et al., 2008). OsNIP1;1, OsNIP2;2 (OsLsi6), OsNIP3;1, OsNIP3;2 (Bienert et al., 2008), and OsNIP3;3 (Katsuhara et al., 2014) also play an important role in transportation AsIII in rice roots. OsNIP3;2 (Lsi2) is expressed in the lateral and stele region of primary roots, involved in the xylem loading of AsIII showing efflux of AsIII from the roots (Ma et al., 2008). Another class of AQP such as plasma membrane intrinsic proteins (PIPs) OsPIP2;4, OsPIP2;6, and OsPIP2;7 are also major players in AsIII uptake and transport in rice (Mosa et al., 2012). Further, the role of NRAMP1 (natural resistance-associated macrophage protein 1) has been identified in the uptake and transport of AsIII (Tiwari et al., 2014).

In the environment, the organic species of As such as monomethyl arsenic acid (MMA), and dimethylarsinic acid (DMA) also exist. The presence of organic species is enhanced by anthropogenic activities such as the application of As-containing herbicides and pesticides. Microorganisms present in the soil are also responsible for the biomethylation of inorganic As species (Huang, 2014). Rice grain accumulates around two times higher methylated forms of As in comparison to Asi (Zheng et al., 2013). However, reports are available for the uptake and transport of methylated forms of As in plants. OsLsi1 is responsible for methylated As species (MMA and DMA) in rice roots from soil (Mawia et al., 2020). Rice plants with lower expression of OsLsi1 or lacking expression of OsLsi1 have been shown to reduce the As (DMA and MMA) uptake and accumulation in comparison with wild type when grown in As-supplemented media. However, OsLsi2 was not found to transport the DMA (Mawia et al., 2020).

4.4.2 TRANSLOCATION OF AS FROM ROOT TO SHOOT

As soon as arsenic enters the cells in any form it is detoxified so that the plants can survive. Plants have various detoxification mechanisms to combat As toxicity. Reduction of AsV to AsIII and further its sequestration inside the vacuole is a very important mechanism of As detoxification in plants. Two independent studies done on *A. thaliana* reported new protein

HAC1 (high arsenic content 1) or ATQ1 (arsenate tolerance QTL1) with AsV reductase activity (Chao et al., 2014; Sanchez-Bermejo et al., 2014). Knockout of the line of HAC1 or ARQ1 had shown lesser efflux of As III following AsV uptake and thus resulting in As hyperaccumulation in shoots. Shi et al. (2016) identified two genes OsHAC1;1 and OsHAC1;2, in rice plants that were close homologs of AtHAC1 in *Arabidopsis* and demonstrated that they function as arsenate reductase (AR) in rice having an important role in As accumulation in shoots and grains. AsIII has a very higher affinity with the –SH group containing proteins. AsV or the methylated forms of As cannot bind with the –SH group containing proteins such as phytochelatins. The rice plant species produce higher content of PCs to bind AsIII and then the complex (AsIII-PCs) is sequestered inside the vacuole. Song et al. (2014) had shown the involvement of two ABC proteins (ATP binding cassette) in the sequestration of AsIII-PCs complexes to the vacuole in *A. thaliana*. AtABCC1 and AtABCC2 are localized at the tonoplast and thus directly transport the AsIII-PCs complex to the vacuole. In rice, OsABCC1 is responsible for the sequestration of the AsIII-PCs complex to the vacuole (Song et al., 2014a). AsIII or AsV which are not sequestered inside the vacuole either may be converted to their organic form and all these species whether the organic or inorganic form is then loaded to the xylem for further translocation to above-ground parts. The transporters, responsible for xylem loading are of very importance because via modulating the expression of these transporters, accumulation in above-ground parts of either hyperaccumulator or crop plant can be regulated. In hyperaccumulator plants, As is stored mainly in the above-ground parts. In *P. vittata*, As is transported and stored mainly in leaf fronds. However, in non-hyperaccumulator plants, As is predominantly stored in the roots, and its transport to the above-ground parts is restricted.

In rice, uptake and xylem loading of AsIII share common pathways with Si. Lsi2 was localized at the proximal side of root parenchymatous cells both in exodermis and endodermis and helped in the efflux of Si and AsIII. The mutant lines for Lsi2 had shown a reduced concentration of AsIII in xylem sap followed by rice grain. However, no effect had been noticed in AsIII uptake. It indicated that OsLsi2 is responsible for AsIII efflux and helps in the loading of AsIII in the xylem (Ma et al., 2008). Along with OsLsi2, OsLsi6 has also a role in the distribution of AsIII. OsLsi6 has mainly expressed in the xylem parenchymatous cells of shoots in rice plants (Detmann et al., 2012). Further, OsNRAMP1 has been shown to have a role in xylem mediated As loading and translocation. Over-expression of OsNRAMP1 in *Arabidopsis* resulted in increased tolerance against AsIII and AsV stress (Tiwari et al., 2014). AsV is rarely translocated from root to shoot because most of the AsV that enters the cells is reduced to AsIII. However, the Pi transporters such as OsPht1;8 have been shown to transport AsV from root to shoot. The over-expressing lines of OsPht1;8 had a higher ability to translocate AsV from roots to shoot (Wu et al., 2011). OsPT2 which is mainly responsible for transferring Pi from roots to shoots may also participate in AsV translocation from root to shoot (Yang et al., 2018).

AsIII and AsV are more efficiently uptaken by the roots of plants in comparison to methylated forms. However, the methylated form of As is prominently present in xylem sap and even in the grain of rice plants. Transport of As from roots to shoots increases in proportion to increases in methylation of As (Zhao et al., 2013). In comparison to MMA, DMA is enormously mobile through the xylem and phloem in rice. OsPTR7 may interrelate in the translocation of DMA in the phloem or xylem. High mobility of DMA occurs in the moving from roots to shoots and from leaves to grain (Tang et al., 2017).

4.4.3 TRANSLOCATION FROM SHOOT TO LEAVES AND GRAIN

Rice is a main staple food crop that is consumed by more than half of the human population. Accumulation of As in rice grain is a serious threat to human health. Rice is used in baby food products also. Consequently, it is very important to understand the mechanisms and the transporters that play an important role in transferring or translocating the As from the shoot to the rice grain.

Nodes of rice plants act as barriers to entry of As in rice grains because of their continuous connection with upper and lower nodes. Most of the time the accumulation of As is higher in nodes in comparison to internodes thus is important in regulating its distribution to rice grain (Moore et al., 2014). Arsenic is mainly transferred from shoot to grain via phloem loading. However, different forms of As are differentially transferred and stored in the rice grain. Carey et al. (2014) demonstrated that phloem loading is involved in the translocation of 55% of DMA and 90% of the AsIII stored in rice grain. It indicated that the phloem loading and unloading is responsible for AsIII accumulation in rice grain. However, both xylem and phloem loading and unloading are equally important for DMA accumulation in rice grain. Phloem transport is accountable for 55% of DMA, 56% of MMA, 100% of AsV, and 90% of AsIII. Translocation of DMA occurs more efficiently in rice plants due to a lack of complexation in molecules. High pH in phloem sap makes the AsIII-PCs complex unstable and thus AsIII may also be present in the phloem sap. The re-translocation of DMA and MMA from the flag leaf into the filling grain is more efficient than AsV and AsIII, which are not detectable even when leaves were fed with AsIII. Duan et al. (2016) identified AtINT2 and AtINT4 (inositol transporters) play important role in phloem loading of AsIII and accumulation in seeds. In rice, OsPTR7 is taking part in long-distance transport and accumulation of DMA in rice grains (Tang et al., 2017).

Although few reports are available for the xylem transport of As, there is scanty information about the mechanisms of phloem-mediated transport of As to grains. There is a clear-cut Synchrotron analysis that revealed a differential pattern in the distribution of As species in the rice grain. AsIII is predominantly engaged in the ovular vascular trace (OVT) region of the rice grain, which is the vascular entrance point into the rice grain and encompasses phloem and xylem cells. However, DMA is present throughout the outer layer as well as in the endosperm of rice grain. It is highlighted that DMA is derived from the re-translocation of DMA in rice grain and accumulated before flowering. However, the Asi mainly transported rice grains during flowering. In this respect, DMAV may accumulate at considerable concentrations in rice grains during the initial stage of grain filling and later diluted by carbohydrate filling the grain (Kumarathilaka et al., 2018).

4.5 CONCLUSION AND FUTURE PROSPECTIVE

Arsenic contamination in rice is hazardous not only for human welfare but also for animals that depend on the above-ground parts for their grazing and food requirements. A few methods such as agronomical practices as well as biological practices may be used to dilute this problem in the case of rice. In the areas where the water is highly contaminated with As alternative water management practices should be adopted to minimize As accumulation in rice grain. The amendments of Si and Pi could also be done and checked the fertilizers usage before application to the soil. Bioremediation practices using microbes involved in the methylation of

As might be helpful in decreasing the bioavailability and uptake of As. The further transgenic approach could also be helpful in decreasing the As content by modulating the expression of a few key genes responsible for As uptake, translocation, and accumulation. It is important to elucidate whether changes in the expression of As transporters may influence the uptake and translocation of other essential elements and compounds in rice plants. However, more rigorous research is indispensable to ascertain the promising mechanism that can be used to reduce As toxicity and accumulation in rice. These analyzes can diminish the unwarranted use of far-off chemicals or compounds on plants that unconstructively prejudiced the environment by destructive plants and patrons directly or indirectly. In adding up, not as much attention has been bountiful to resolving the crisis of drinking water globally and irrigating crops with As contaminated water which finally resulted in the arsenicosis epidemic and malnutrition in many developing countries. Consequently, it is an essential requirement for any technique that can be utilized to extract As or remediate As-contaminated water before utilizing it for irrigation or drinking purposes.

ACKNOWLEDGMENT

RDT and PKS are thankful to the National Academy of Sciences, Prayagraj, India for the NASI-Senior Scientist Platinum Jubilee fellowship and RAship, respectively. In addition, SD acknowledges SEED-DST, New Delhi for the project grant (SEED/TIASN/2018/74).

KEYWORDS

- **arsenic**
- **class 1 human carcinogen**
- **monomethylarsonic acid**
- **non-essential toxic element**
- ***Oryza sativa***
- **rice grain**
- **transporter**

REFERENCES

Abedin, M. J., Feldmann, J., & Meharg, A. A., (2002). Uptake kinetics of arsenic species in rice plants. *Plant Physiology, 128*(3), 1120–1128.

Azam, S. M. G. G., Sarker, T. C., & Naz, S., (2016). Factors affecting the soil arsenic bioavailability, accumulation in rice and risk to human health: A review. *Toxicology Mechanisms and Methods, 26*(8), 565–579.

Baffes, J., & Gautam, M., (2001). Assessing the sustainability of rice production growth in Bangladesh. *Food Policy, 26*(5), 515–542.

Bissen, M., & Frimmel, F. H., (2003). Arsenic-a review. Part I: Occurrence, toxicity, speciation, mobility. *Acta Hydrochimica et Hydrobiologica, 31*(1), 9–18.

Chakraborty, M., Mukherjee, A., & Ahmed, K. M., (2015). A review of groundwater arsenic in the Bengal Basin, Bangladesh and India: From source to sink. *Current Pollution Reports, 1*(4), 220–247.

Charlet, L., Chakraborty, S., Appelo, C. A. J., Roman-Ross, G., Nath, B., Ansari, A. A., Lanson, M., et al., (2007). Chemodynamics of an arsenic "hotspot" in a West Bengal aquifer: A field and reactive transport modeling study. *Applied Geochemistry, 22*(7), 1273–1292.

Chen, Z., Zhu, Y. G., Liu, W. J., & Meharg, A. A., (2005). Direct evidence showing the effect of root surface iron plaque on arsenite and arsenate uptake into rice (*Oryza sativa*) roots. *New Phytologist*, 91–97.

Deschamps, E., Ciminelli, V. S., Weidler, P. G., & Ramos, A. Y., (2003). Arsenic sorption onto soils enriched in Mn and Fe minerals. *Clays and Clay Minerals, 51*(2), 197–204.

Du Laing, G., Meers, E., Dewispelaere, M., Rinklebe, J., Vandecasteele, B., Verloo, M. G., & Tack, F. M., (2009). Effect of water table level on metal mobility at different depths in wetland soils of the Scheldt estuary (Belgium). *Water, Air, and Soil Pollution, 202*(1), 353–367.

Gersztyn, L., Karczewska, A., & Gałka, B., (2013). Influence of pH on the solubility of arsenic in heavily contaminated soils. *Environmental Protection and Natural Resources, 24*(3), 7–11.

Ghosh, D., Bhadury, P., & Routh, J., (2014). Diversity of arsenite oxidizing bacterial communities in arsenic-rich deltaic aquifers in West Bengal, India. *Frontiers in Microbiology, 5*, 602.

Ghosh, N. C., & Singh, R. D., (2009). Groundwater arsenic contamination in India: Vulnerability and scope for remedy. In: *Technical Paper Included in the Special Session on Ground Water in the 5th Asian Regional Conference of INCID* (pp. 9–11).

Goldberg, S., (2002). Competitive adsorption of arsenate and arsenite on oxides and clay minerals. *Soil Science Society of America Journal, 66*(2), 413–421.

Hossain, M. B., Jahiruddin, M., Panaullah, G. M., Loeppert, R. H., Islam, M. R., & Duxbury, J. M., (2008). Spatial variability of arsenic concentration in soils and plants, and its relationship with iron, manganese, and phosphorus. *Environmental Pollution, 156*(3), 739–744.

Hussain, M. M., Bibi, I., Niazi, N. K., Shahid, M., Iqbal, J., Shakoor, M. B., Ahmad, A., et al., (2021). Arsenic biogeochemical cycling in paddy soil-rice system: Interaction with various factors, amendments, and mineral nutrients. *Science of the Total Environment, 773*, 145040.

IARC Working Group on the Evaluation of Carcinogenic Risks to Humans, (2004). Some drinking-water disinfectants and contaminants, including arsenic. In: *IARC Monographs on the Evaluation of Carcinogenic Risks to Humans* (Vol. 84, pp. 1–477).

Isayenkov, S. V., & Maathuis, F. J., (2008). The *Arabidopsis thaliana* aquaglyceroporin AtNIP7; 1 is a pathway for arsenite uptake. *FEBS Letters, 582*(11), 1625–1628.

Jang, Y. C., Somanna, Y., & Kim, H. J. I. J., (2016). Source, distribution, toxicity and remediation of arsenic in the environment: A review. *Int. J. Appl. Environ. Sci., 11*(2), 559–581.

Juhasz, A. L., Smith, E., Weber, J., Rees, M., Rofe, A., Kuchel, T., Sansom, L., & Naidu, R., (2006). *In vivo* assessment of arsenic bioavailability in rice and its significance for human health risk assessment. *Environmental Health Perspectives, 114*(12), 1826–1831.

Khan, M. A., Stroud, J. L., Zhu, Y. G., McGrath, S. P., & Zhao, F. J., (2010). Arsenic bioavailability to rice is elevated in Bangladeshi paddy soils. *Environmental Science & Technology, 44*(22), 8515–8521.

Kumpiene, J., Lagerkvist, A., & Maurice, C., (2008). Stabilization of As, Cr, Cu, Pb and Zn in soil using amendments: A review. *Waste Management, 28*(1), 215–225.

Kuramata, M., Sakakibara, F., Kataoka, R., Abe, T., Asano, M., Baba, K., Takagi, K., & Ishikawa, S., (2015). Arsenic biotransformation by *S treptomyces* sp. isolated from rice rhizosphere. *Environmental Microbiology, 17*(6), 1897–1909.

Li, H., Chen, X. W., & Wong, M. H., (2016). Arbuscular mycorrhizal fungi reduced the ratios of inorganic/organic arsenic in rice grains. *Chemosphere, 145*, 224–230.

Li, R. Y., Ago, Y., Liu, W. J., Mitani, N., Feldmann, J., McGrath, S. P., Ma, J. F., & Zhao, F. J., (2009). The rice aquaporin Lsi1 mediates uptake of methylated arsenic species. *Plant Physiology, 150*(4), 2071–2080.

Li, X., Hu, Y., Gong, J., Zhang, L., & Wang, G., (2013). Comparative genome characterization of *Achromobacter* members reveals potential genetic determinants facilitating the adaptation to a pathogenic lifestyle. *Applied Microbiology and Biotechnology, 97*(14), 6413–6425.

Liu, W. J., Zhu, Y. G., Hu, Y., Williams, P. N., Gault, A. G., Meharg, A. A., Charnock, J. M., & Smith, F. A., (2006). Arsenic sequestration in iron plaque, its accumulation and speciation in mature rice plants (*Oryza sativa* L.). *Environmental Science & Technology, 40*(18), 5730–5736.

Liu, W. J., Zhu, Y. G., Smith, F. A., & Smith, S. E., (2004). Do phosphorus nutrition and iron plaque alter arsenate (As) uptake by rice seedlings in hydroponic culture? *New Phytologist, 162*(2), 481–488.

Lockwood, C. L., Mortimer, R. J., Stewart, D. I., Mayes, W. M., Peacock, C. L., Polya, D. A., Lythgoe, P. R., et al., (2014). Mobilization of arsenic from bauxite residue (red mud) affected soils: Effect of pH and redox conditions. *Applied Geochemistry, 51*, 268–277.

Ma, J. F., Yamaji, N., Mitani, N., Xu, X. Y., Su, Y. H., McGrath, S. P., & Zhao, F. J., (2008). Transporters of arsenite in rice and their role in arsenic accumulation in rice grain. *Proceedings of the National Academy of Sciences, 105*(29), 9931–9935.

Mangalgiri, K. P., Adak, A., & Blaney, L., (2015). Organoarsenicals in poultry litter: Detection, fate, and toxicity. *Environment International, 75*, 68–80.

Martin, S. T., (2005). Precipitation and dissolution of iron and manganese oxides. *Environmental Catalysis, 1*, 61–82.
Meharg, A. A., & Macnair, M. R., (1992). Suppression of the high affinity phosphate uptake system: A mechanism of arsenate tolerance in *Holcus lanatus* L. *Journal of Experimental Botany, 43*(4), 519–524.
Meharg, A. A., (2004). Arsenic in rice–understanding a new disaster for South-East Asia. *Trends in Plant Science, 9*(9), 415–417.
Mondal, D., & Polya, D. A., (2008). Rice is a major exposure route for arsenic in Chakdaha block, Nadia district, West Bengal, India: A probabilistic risk assessment. *Applied Geochemistry, 23*(11), 2987–2998.
Mondal, P., Majumder, C. B., & Mohanty, B., (2006). Laboratory based approaches for arsenic remediation from contaminated water: Recent developments. *Journal of Hazardous Materials, 137*(1), 464–479.
Nriagu, J. O., Bhattacharya, P., Mukherjee, A. B., Bundschuh, J., Zevenhoven, R., & Loeppert, R. H., (2007). Arsenic in soil and groundwater: An overview. *Trace Metals and Other Contaminants in the Environment, 9*, 3–60.
Onken, B. M., & Hossner, L. R., (1995). Plant uptake and determination of arsenic species in soil solution under flooded conditions. *American Society of Agronomy, Crop Science Society of America, and Soil Science Society of America, 24*(2), 373–381.
Ravenscroft, P., Brammer, H., & Richards, K., (2011). *Arsenic Pollution: A Global Synthesis* (Vol. 94). John Wiley & Sons.
Romero-Freire, A., Peinado, F. J., Ortiz, M. D., & Van, G. C. A. M., (2015). Influence of soil properties on the bioaccumulation and effects of arsenic in the earthworm *Eisenia andrei*. *Environmental Science and Pollution Research, 22*(19), 15016–15028.
Santini, J. M., Sly, L. I., Schnagl, R. D., & Macy, J. M., (2000). A new chemolithoautotrophic arsenite-oxidizing bacterium isolated from a gold mine: Phylogenetic, physiological, and preliminary biochemical studies. *Applied and Environmental Microbiology, 66*(1), 92–97.
Shaji, E., Santosh, M., Sarath, K. V., Prakash, P., Deepchand, V., & Divya, B. V., (2021). Arsenic contamination of groundwater: A global synopsis with focus on the Indian Peninsula. *Geoscience Frontiers, 12*(3), 101079.
Sharma, S., Anand, G., Singh, N., & Kapoor, R., (2017). Arbuscular mycorrhiza augments arsenic tolerance in wheat (*Triticum aestivum* L.) by strengthening antioxidant defense system and thiol metabolism. *Frontiers in Plant Science, 8*, 906.
Shi, K., Wang, Q., & Wang, G., (2020). Microbial oxidation of arsenite: Regulation, chemotaxis, phosphate metabolism and energy generation. *Frontiers in Microbiology, 11*, 2235.
Shrivastava, A., Ghosh, D., Dash, A., & Bose, S., (2015). Arsenic contamination in soil and sediment in India: Sources, effects, and remediation. *Current Pollution Reports, 1*(1), 35–46.
Straw, B., Dewey, C., Kober, J., & Henry, S. C., (2002). Factors associated with death due to hemorrhagic bowel syndrome in two large commercial swine farms. *Journal of Swine Health and Production, 10*(2), 75–79.
Stroud, J. L., Norton, G. J., Islam, M. R., Dasgupta, T., White, R. P., Price, A. H., Meharg, A. A., et al., (2011). The dynamics of arsenic in four paddy fields in the Bengal delta. *Environmental Pollution, 159*(4), 947–953.

Száková, J., Tlustoš, P., Goessler, W., Frková, Z., & Najmanová, J., (2009). Mobility of arsenic and its compounds in soil and soil solution: The effect of soil pretreatment and extraction methods. *Journal of Hazardous Materials, 172*(2, 3), 1244–1251.

Tripathi, R. D., Srivastava, S., Mishra, S., Singh, N., Tuli, R., Gupta, D. K., & Maathuis, F. J., (2007). Arsenic hazards: Strategies for tolerance and remediation by plants. *Trends in Biotechnology, 25*(4), 158–165.

Tripathi, R. D., Tripathi, P., Dwivedi, S., Kumar, A., Mishra, A., Chauhan, P. S., Norton, G. J., & Nautiyal, C. S., (2014). Roles for root iron plaque in sequestration and uptake of heavy metals and metalloids in aquatic and wetland plants. *Metallomics, 6*(10), 1789–1800.

Tuli, R., Chakrabarty, D., Trivedi, P. K., & Tripathi, R. D., (2010). Recent advances in arsenic accumulation and metabolism in rice. *Molecular Breeding, 26*(2), 307–323.

Wang, H. L., Hu, Z. H., Tong, Z. L., Xu, Q., Wang, W., & Yuan, S., (2014). Effect of arsanilic acid on anaerobic methanogenic process: Kinetics, inhibition and biotransformation analysis. *Biochemical Engineering Journal, 91*, 179–185.

Wang, Q., Warelow, T. P., Kang, Y. S., Romano, C., Osborne, T. H., Lehr, C. R., Bothner, B., et al., (2015). Arsenite oxidase also functions as an antimonite oxidase. *Applied and Environmental Microbiology, 81*(6), 1959–1965.

Williams, P. N., Price, A. H., Raab, A., Hossain, S. A., Feldmann, J., & Meharg, A. A., (2005). Variation in arsenic speciation and concentration in paddy rice related to dietary exposure. *Environmental Science & Technology, 39*(15), 5531–5540.

World Health Organization, (1993). *Guidelines for Drinking-Water Quality* (2nd edn., Vol. 1). Geneva.

Xu, X. Y., McGrath, S. P., Meharg, A. A., & Zhao, F. J., (2008). Growing rice aerobically markedly decreases arsenic accumulation. *Environmental Science & Technology, 42*(15), 5574–5579.

Yamaguchi, N., Nakamura, T., Dong, D., Takahashi, Y., Amachi, S., & Makino, T., (2011). Arsenic release from flooded paddy soils is influenced by speciation, Eh, pH, and iron dissolution. *Chemosphere, 83*(7), 925–932.

Yao, L., Huang, L., He, Z., Zhou, C., & Li, G., (2013). Occurrence of arsenic impurities in organoarsenics and animal feeds. *Journal of Agricultural and Food Chemistry, 61*(2), 320–324.

Yeasmin, T., Zaman, P., Rahman, A., Absar, N., & Khanum, N. S., (2007). Arbuscular mycorrhizal fungus inoculum production in rice plants. *African Journal of Agricultural Research, 2*(9), 463–467.

Zavala, Y. J., & Duxbury, J. M., (2008). Arsenic in rice: I. Estimating normal levels of total arsenic in rice grain. *Environmental Science & Technology, 42*(10), 3856–3860.

Zhao, F. J., Ma, J. F., Meharg, A. A., & McGrath, S. P., (2009). Arsenic uptake and metabolism in plants. *New Phytologist, 181*(4), 777–794.

CHAPTER 5

Arsenic-Induced Phytotoxicity in Rice Plants

PRASHANT SINGH,[1] NANDINI GUPTA,[2] PAYAL M. DEOGHARE,[1] and RAKHI AGARWAL[2]

[1]*School of Medico-Legal Studies, National Forensic Sciences University, India*

[2]*School of Forensic Sciences, National Forensic Sciences University, India*

ABSTRACT

Arsenic in rice is one of the oldest pollutants, contaminants, and poisons that has caused a lot of trouble to humans and the environment. Arsenic being a highly toxic heavy metal exhibits harmful effects that are known to even cause death. After industrialization, a lot of processes have traces of arsenic that led to its bioaccumulation in the environment leading to its persistence in the widely consumed crop "rice." Rice is one of the largest consumed foods in the world that has been seriously affected by arsenic contamination. Arsenic after accumulating in rice plants is known to induce phytotoxicity through various mechanisms which pose a threat to both the rice plant and the consumer. Moreover, it disturbs the natural environment surrounding it. This chapter discusses the issue of arsenic-induced phytotoxicity in rice plants, followed by the various mechanisms involved in causing the damage. This chapter also overviews why such phytotoxicity persists in rice plants and the potential sources of arsenic contamination in the rice plants.

Arsenic in Rice: A Global Menace. Garima Awasthi, Sudhakar Shrivastava, and Mahipal Singh Sankhla, (Eds.)

5.1 INTRODUCTION

Arsenic is one of the most notorious environmental pollutants that pose high toxicity risks in animals and plants (Bhattacharya et al., 2010). WHO (World Health Organization) has recognized arsenic as a carcinogen metalloid (Shi et al., 2014). Various studies indicated that the consumption of arsenic-contaminated rice causes serious health problems (Bhattacharya et al., 2010). Naturally, arsenic exists in inorganic and organic forms in 4 different oxidation states –3, 0, +3, +5 (Patel et al., 2005).

Apart from the various sources mentioned in Table 5.1, the major accumulation of arsenic in rice is facilitated by contaminated groundwater (Patel et al., 2005). Toxicity from arsenic is associated with cancer, hyperkeratosis, hypertension, gangrene, black foot disease, melanosis, restrictive lung disease, and diabetes (Bhattacharya et al., 2010; Das et al., 2004).

TABLE 5.1 Discussing the Various Sources of Arsenic in Rice Crops

Sources of Arsenic in Rice
• From pesticides, herbicides, and chemical fertilizers in agriculture and forestry.
• As a feed additive in the glass/ceramic industry.
• As an alloying agent.
• In manufacturing agents of pharmaceuticals, semiconductors, explosives, etc.
• Anthropogenic activities like mining, smelting, farming, etc.
• Contamination in the environment from natural sources like rock weathering, and deposition from the atmosphere.
• Irrigation of food crops (like rice) with arsenic-contaminated groundwater.

Soil and water contaminated with arsenic are responsible for contaminated crops (Williams et al., 2005). One such example is rice, one of the largest foods consumed around the world that has been reported for arsenic contamination (Meacher et al., 2002; Tsuji et al., 2007). Most South Asian countries are reported to be affected by arsenic and its related phytotoxicity due to contamination of arsenic in groundwater (Chowdhury et al., 2000; Smith et al., 2000; Pandey et al., 2002) in which rice was grown. It has been found that the concentration of arsenic in rice is present ten-fold (Meharg & Zhao, 2012). Rice is more prone to arsenic contamination and accumulation, as it is grown in more water manifested areas (Hojsak et al., 2015).

Arsenic is known to induce phytotoxicity in rice plants by various mechanisms which pose huge health risks (Biswas et al., 2020). Hence, it becomes essential to understand the phytotoxicity that arsenic causes to rice plants by various mechanisms that are discussed in this chapter. A flowchart of arsenic-induced phytotoxicity to rice plants is mentioned in Figure 5.1.

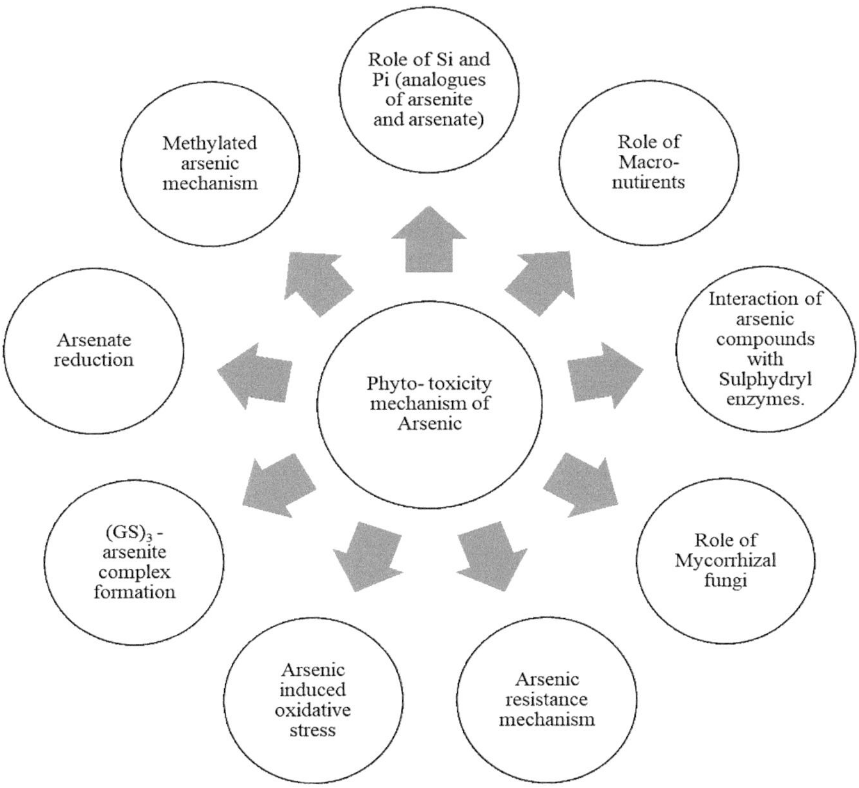

FIGURE 5.1 Summary of arsenic-induced phytotoxicity in rice plants.

5.2 VARIOUS MECHANISMS OF ARSENIC PHYTOTOXICITY IN RICE PLANTS

5.2.1 ARSENATE REDUCTION MECHANISM IN RICE PLANT

Various forms of arsenic are found in plant tissues such as As(III), As V, and various organic Arsenals. Majorly, arsenic in rice is present in the form of Arsenite (As III) even if the plant is exposed to Arsenate (As V).

The mechanism of arsenate reduction involves three types of cytosolic arsenate reductases (ArsC). Firstly, *E. coli* ArsC utilizes glutaredoxin (Grx) and glutathione (GSH) as reductants. During this non-enzymatic process, arsenate is reduced to arsenite as an outcome of the oxidation of two reduced glutathione molecules to form oxidized GSH which is immediately recycled into 2 GSH molecules by GSH reductase (Mawia et al., 2021). Secondly, ArsC uses thioredoxin as a reductant, which was identified from the *Staphylococcus aureus* and *Bacillus subtilis.* Thirdly, ACR2p which utilizes Grx and GSH as reductants were identified from yeast and parasitic protozoa (F. J. Zhao et al., 2009).

ACR2p is a part of the protein tyrosine phosphatase which involves human cell cycle dual-specificity phosphatases CDC25s (cell division cycle). Extracted Acr2 genes from rice and other plants also act similarly to CDC25s-like genes along with two homologs of rice OsACR2 (OsACR2;1 and OsACR2;2) that perform like enzyme phosphatase. In *E. coli,* OsACR gene expression was observed and it showed the reversal of arsenate reduction to arsenite, stipulating the involvement of OsACR in the reduction reaction of Arsenate (As V) to Arsenite (As III) in rice plants (Majumdar & Bose, 2017).

5.2.2 METHYLATED ARSENIC MECHANISM OF PHYTOTOXICITY IN RICE

The two forms of inorganic arsenic (Asi) are; trivalent arsenic As^{3+} and pentavalent arsenic As^{5+}; out of these two, trivalent arsenic is more toxic in rice plants. The microorganisms present in the soil convert the Asi into methylated form resulting in the formation of monomethyl arsenic acid (MMAsv) and dimethyl arsenic acid (DMAsv) (Rehman & Naranmandura, 2012) MMA and DMA are the final metabolites of the reductive methylation pathway of arsenic in rice. Along with these two methylated forms, Trimethyl arsine (TMAs) is also present in soil environments in which rice plants are grown (Rehman & Naranmandura, 2012).

Rice plants majorly contain dimethyl arsenic acid (DMAsv), minorly monomethyl arsenic acid (MMAsv), trimethyl arsine oxide (TMAsvO), and rarely tetramethyl arsonium cation (TETRAsv).

In the Methylation of arsenite (AsIII) in rice, firstly, pentavalent arsenic (AsV) is reduced to Trivalent arsenic (AsIII), then methyl is transferred from S-adenosyl-L-methionine (SAM) to AsIII by oxidative methylation reaction and is catalyzed by the enzyme AsIII S-adenosylmethionine methyltransferase (ArsM) (Rehman & Naranmandura, 2012). This reaction further results in the formation of pentavalent MMAs (V), which get reduced to trivalent MMAs (III) in the presence of reductase. MMAs (III) further get reduced to DMAs (III) and lastly to volatile TMAs. Bacteria, fungi, and archaea are homologous to this enzyme and also possess the ability to methylate the Asi species in rice, As III into volatile arsines (Jia et al., 2013) (Figure 5.2).

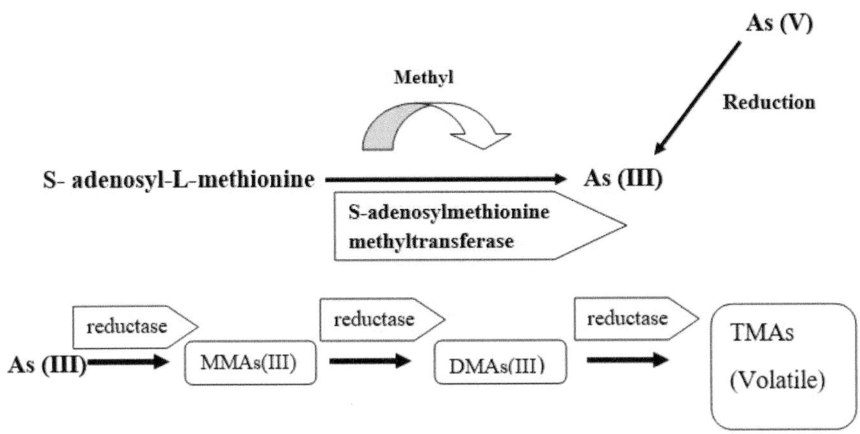

FIGURE 5.2 The methylated arsenic mechanism in arsenic-contaminated rice plants.

Rice plants uptake arsenic species and methylated arsenic species through roots and shoots in pentavalent form (MMA(V) and DMA (V)) which further gets reduced to trivalent form (MMA(III) and DMA(III)) in the rice plant.

5.2.3 (GS)$_3$-ARSENITE COMPLEX FOR ARSENITE-INDUCED PHYTOTOXICITY

There are two steps for the synthesis of GSH in rice plants. After arsenic uptake by the rice plants, First, γ-glutamyl-cysteine is produced in the reaction which is dependent on ATP and catalyzed by glutamate-cysteine ligase which is also called γ-glutamyl-cysteine synthetase. The addition of glycine to γ-glutamyl-cysteine is catalyzed by Glutathione synthetase (Zagorchev et al., 2013) (Figure 5.3).

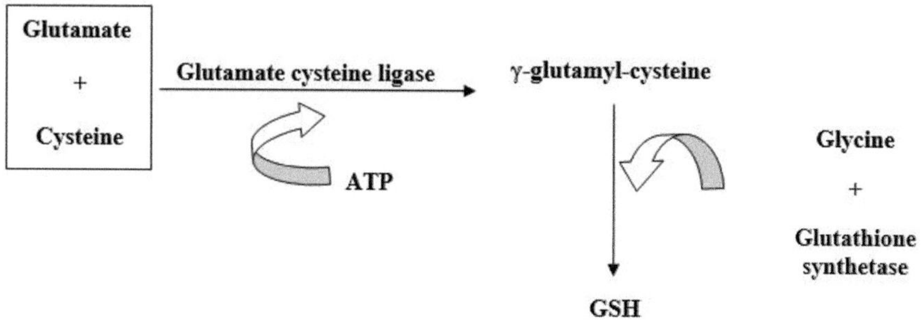

FIGURE 5.3 Synthesis of GSH in rice plants contaminated with arsenic.

Two Sulfhydryl groups (-SH) of peptides like GSH and Phytochelatin have a greater affinity to bind with arsenic toxic forms. GSH and arsenite combine to form $(GS)_3$-arsenite complex with cysteinyl sulfhydryl because of the arsenite binding site. The resulting complex ($(GS)_3$-arsenite) remains stable at 1.5 to 7.5 pH but breaks down when the pH exceeds (Zagorchev et al., 2013) (Figure 5.4).

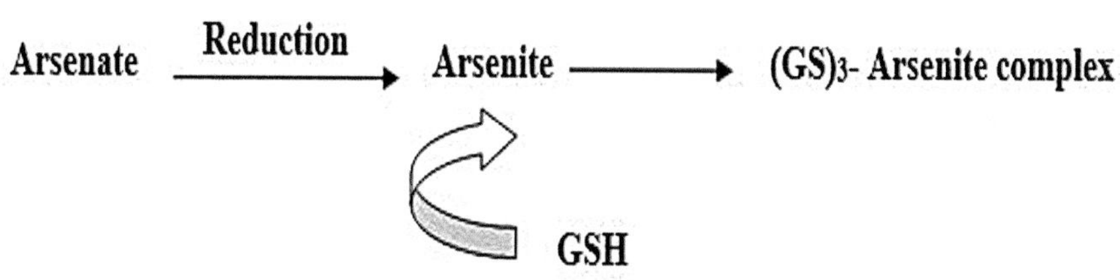

FIGURE 5.4 Formation of $(GS)_3$-arsenite complex in rice plants.

It is thought that the toxicity of the arsenite in rice plants is due to the binding of arsenite to the –SH groups of proteins, resulting in changes in the structure of protein and causing a disturbance in the catalytic sites of enzymes. Organic arsenate, i.e., arsenate in DMA can bind with GSH when it is operated by sulfide (Zagorchev et al., 2013).

Arsenic detoxification can be done with the help of the complexation of arsenite with Phytochelatin (PCs) (Zagorchev et al., 2013). When the rice plant is exposed to arsenate or arsenite, it causes increased synthesis and deposition of Phytochelatin. Phytochelatin-Arsenic complexes are found in intact form within the rice plant. Along with the Phytochelatin-Arsenic complex, certain other complexes are also identified, such as As(III)-(PC2)2, As(III)-PC3, GS-As(III)-PC2, and MMA-PC2, in the rice plants (Zhao et al., 2009).

5.2.4 ARSENIC-INDUCED OXIDATIVE STRESS IN RICE PLANTS

Arsenic is known to induce toxic Reactive Oxygen Species (ROS) in rice plants which are generated in the cell wall region as well as inside the cell. These ROS produced within the rice plants affects the electron transport chain, membrane permeability, enzyme activity, metabolic pool, leaf chlorosis, necrosis, and ion homeostasis that leads to a decrease in rice plant's biomass (Gaikwad et al., 2020). Studies showed that arsenic-induced phytotoxicity via ROS mechanisms is arising due to the presence of arsenic in the soil and/or irrigated groundwater, where rice plants are grown (Gaikwad et al., 2020). At the cellular level, arsenic not only damages the chloroplast membrane but also alters the functions of photosynthetic systems and serves as the site for ROS generation in rice plants. The imbalance between ROS production and antioxidative system response is the direct cause of oxidative stress in rice (Gaikwad et al., 2020).

In order to protect rice from the sudden increase in intracellular levels of ROS, rice cells contain both enzymatic antioxidants, superoxide dismutase (SOD), catalase (CAT), ascorbate peroxidase (APX), guaiacol peroxidase (GPOX), monodehydroascorbate reductase (MDHAR), dehydroascorbate reductase (DHAR) and glutathione reductase (GR), as well as non-enzymatic antioxidants, such as ascorbate, glutathione, carotenoid, α-tocopherol and accumulation of anthocyanin in the leaves, have been found to be very important (Gaikwad et al., 2020). The key role is played by the SOD enzyme which belongs to a group of metalloenzymes and catalyzes the dismutation of superoxide free radicals (O_2^-) into O_2 and H_2O_2. Higher activity of SOD may be the cause of this H_2O_2 production during the arsenic stress in rice. The increase in APX, SOD, and POD activities could represent appropriate protection against the overproduction of peroxides when arsenic is accumulated in rice. It was observed from the study conducted by Gaikwad et al. that the rice crop was found to be more sensitive to As III as compared to As V (Gaikwad et al., 2020).

Various inorganic species of arsenic, i.e., Arsenite (As III) and arsenate species (As V) are found in the soil in which rice was grown. The effect of arsenic toxicity in the rice plants is due to the generation of Reactive Oxygen Species (ROS) such as superoxide radicals, hydroxyl radicals, and hydrogen peroxide that disrupt the important biomolecules of the rice plants. After entry into the rice plant, arsenic interacts with the cells of the rice plant and causes various physiological and morphological abnormalities (Kumar et al., 2020). Radical Oxygen Loss (ROL) is the radical movement of oxygen from the aerenchyma of roots to the soil which causes the oxidation of Fe^{2+} to Fe^{3+} that results in the formation of iron plaque. Iron plaque is involved in the adsorption and co-precipitation of arsenic in rice as it helps in reducing arsenic uptake and accumulation. ROL is required to control iron plaque formation in rice as it restricts the mobility and bioavailability of the rice plant and helps in reducing arsenic transport and accumulation in rice grains (Shri et al., 2019).

Rice plants release oxygen through their roots, and this leads to iron plaque formation on the root surface. Iron plaque can act as a major sink or source of arsenic to the rice plants. Hence, water management of rice fields can also modify arsenic bioavailability in rice plants through changes in oxygen release from rice plants and subsequent iron plaque formation on rice roots (Upadhyay et al., 2020). When the rice is cultivated, the field is filled with stagnant water, thereby creating anaerobic conditions, and arsenic release from the dissolution of iron oxyhydroxides is promoted. As III remains more attached to the roots of rice than As V and it is released in the soil by microbial reduction, alteration in pH, and redox coupling increasing the bioavailability of more soluble As III. So, water management has to be practiced effectively (Upadhyay et al., 2020).

5.2.5 PHYTOCHELATIN MECHANISM IN RICE PLANTS

Heavy metal accumulation in rice plants can cause several abnormalities/diseases in plants, and various defense mechanisms such as generation of thiol via cysteine (Cys), glutathione (GSH), and Phytochelatin (PC) being carried out by the rice plants. Phytochelatin is a thiol-rich cysteine heavy metal-binding peptide that can detoxify heavy metals in rice plants and is synthesized from glutathione (GSH) by the activity of the PC synthase enzyme (Kumar et al., 2020).

A study was conducted by Duan et al. to find out the role of Phytochelatin in detoxifying cellular arsenic through the complexation of arsenite in rice. For this particular study, six different rice genotypes which were previously determined as high-grain arsenic accumulators or low-grain arsenic accumulators were collected (Duan et al., 2011). The total arsenic analysis for the rice seeds was done by first treating the cells with concentrated nitric acid and then performing microwave digestion and secondly, the arsenic concentration was determined with the help of ICP-MS. It was then concluded from the study that the complexation of arsenite with Phytochelatin reduces arsenic mobility from shoots to grains of rice. Hence, the cultivars with more efficient Phytochelatin synthesis will have a lower arsenic translocation efficiency (Duan et al., 2011) (Figure 5.5).

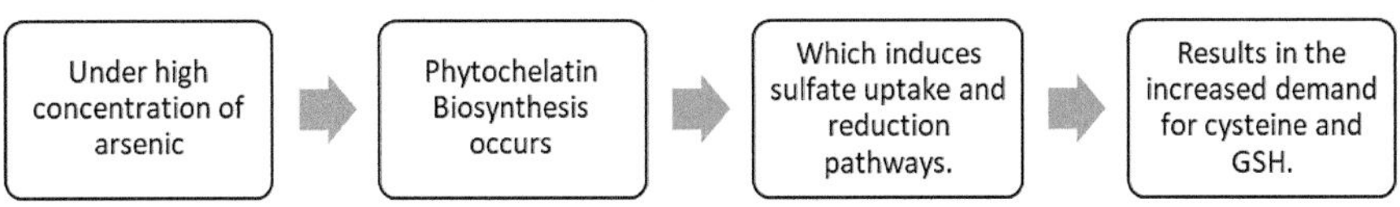

FIGURE 5.5 Increased demand for cysteine and GSH in rice plants.

Rice plants have evolved a homeostasis mechanism to maintain low cellular concentrations of various non-essential elements. Although it is difficult to detect the gross physiological changes in the uptake and metabolism of several micronutrients, internal compartmentalization plays a significant role in the adaptation of the elements. This case does not work with the uptake of non-essential micronutrients such as arsenic by rice plants (Meharg et al., 2012). Arsenate acts as an analog of phosphate. Therefore, rice plants that grow on arsenate-contaminated soils assimilate high levels of arsenate unless they have an altered phosphate transport mechanism. To maintain cellular function, the cytoplasmic concentrations of the arsenate must be kept low. It is shown in the study conducted by Meharg & Macnair (1991, 1992) that arsenate sensitivity in rice is highly linked with phosphate nutrition which indicated that non-resistant rice plants can be made more resistant to arsenate by raising their phosphorus levels. It is indicated in the study conducted by (Fitter et al., 1998; Wright et al., 2000) that arsenate-resistant rice plants differ from nonresistant phenotypes in their reproductive and flowering behavior as they are not able to properly allocate the phosphates (Meharg et al., 2012).

There is a high uptake of silicon and phosphorus in the rice plant which is required for its growth. As III and As V have similar molecular properties to Si and Pi resulting in a higher accumulation of arsenic in rice. OsPT8, and PvPht which are Pi transporter are used for transporting AsV inside the rice plant and show toxic effects. Nodulin 26-like intrinsic protein (NIPs) transports AsIII along with its methylated form (Meharg et al., 2012). Other transporters of As III are PvTIP4;1 which is responsible for the arsenic influx in rice plants. If the transporters of Si and Pi are knocked out, then it significantly reduces the uptake of As V and As III, respectively. In rice, Chen et al. revealed the participation of Lsi2 and Phytochelatin levels during the transport of arsenic to the panicle node and grain of rice (Meharg et al., 2012).

Detoxification of arsenic in rice occurs when it is converted from As V to As III in the presence of AR. In rice, OsHAC1;1 and OsHAC1;2 have been shown to function as As V reductase (Shri et al., 2019).

$$As\,V \xrightarrow{\textit{Arsenate Reductase}} As\,III$$

5.2.6 INTERACTION OF ARSENIC COMPOUNDS WITH SULFHYDRYL ENZYMES OF RICE

Arsenite chelates with sulfhydryl (–SH) rich protein (GSH, PCs, and MTs) and forms a complex that gets sequestered by vacuolar transporters (Kumar et al., 2020). Arsenate being a phosphate analog interferes with phosphate metabolism (phosphorylation and ATP synthesis) in rice plants while As(III) binds to sulfhydryl groups of proteins affecting their structures and/or catalytic functions. In addition, sulfur metabolism plays a crucial role in tackling arsenic inside the rice plants as sulfhydryl (-SH) containing short peptides like glutathione (GSH) and Phytochelatin (PCs) bind with arsenic and help in its sequestration to vacuoles. In addition, the coordination of various metabolic pathways viz. carbon, nitrogen, and sulfur metabolism are essential to tolerate arsenic stress effectively (Awasthi et al., 2017).

Cysteine, which is the final product of sulfate assimilation and a central metabolite, serves as the sulfur donor for the synthesis of Met, iron-sulfur clusters, CoA, thiol-containing proteins, etc., in rice plants. Cysteine synthesis takes place in two steps.

$$Serine + Acetyl - CoA \xrightarrow{o-acetyl\ serine\ (thiol)\ lyase} O - Acetyl\ Serine \xrightarrow{Hydrogen\ Sulphide} Cysteine$$

It is highly significant in checking the stress activity of rice plants. Unstressed cells have a low concentration of free cysteine and stressed cells have a high concentration of cysteine in rice. Although cysteine is a potent chelator of heavy metal ions, free cysteine can produce toxic OH radicals (Awasthi et al., 2017). Therefore, free cysteine can lead to undesirable loss of sulfur which has a deleterious effect on the stress-coping mechanism of the rice plant.

Methionine can also undergo ROS-mediated oxidation to methionine sulfoxide (MetO), and this can lead to changes in the protein conformation and activity of rice.

$$MetO \xrightarrow{Met-Sufoxide\ reductases} Met$$

Methionine is a substrate for the synthesis of various polyamines which play a crucial role in the stress tolerance of rice plants from drought, and abiotic stresses such as salt, drought, and chilling (Zagorchev et al., 2013).

5.2.7 ROLE OF MYCORRHIZAL FUNGI IN ARSENIC-INDUCED PHYTOTOXICITY

The mutuality between rice plants and arbuscular mycorrhizal fungi (AMF) is beneficial for the target rice plant as this increases the possession of the nutrients, majorly phosphorus. In arbuscular mycorrhizal relation, fungal hyphae creep into the cortical cells of the roots of the target rice plant forming a small tree-shaped design called arbuscules which acts as a site of exchange of nutrients between the target rice plant and the Fungi (He & Lilleskov, 2014). Because of the interaction between mycorrhizal fungi and rice plants, arsenic resistance in the target rice plants is granted.

For this resistive property, certain mechanisms are responsible. First, mycorrhizal grouping causes a decrease in the uptake of arsenate (As V) in rice because the grouping suppresses the high-affinity phosphate transport of the root system of rice.

Secondly, mycorrhizal fungi remove arsenic from rice plants to the external environment which enhances the arsenic resistance in target rice plants (Liu et al., 2014; Zhao et al., 2009). Arsenic-tolerant mycorrhizal fungi improve the As III secretion to the outer environment and decrease the As V absorption in rice.

Thirdly, by strengthening the phosphorus gain and fixing the phosphorus content of the nutrients of the target rice plants, mycorrhizal fungi increase rice growth, leading to a difference in arsenic concentration in rice plants. When the phosphorus amount exceeds the arsenic amount, tolerance to arsenic by rice plants increases. Mycorrhizal fungi may act as an intermediate of arsenic biomethylation and evolve the methylated arsenic in the rhizosphere (Zhao et al., 2013; Meharg, 2002). AMF also plays an important role in arsenic uptake by rice plants. Phosphorus and arsenic to some extent are chemically similar. When they combine with the mycorrhizal fungi, they may increase the arsenic uptake because it basically enhances the phosphorus absorption but because of structure similarity, arsenic also gets absorbed in the target rice plant (He & Lilleskov, 2014).

5.2.8 ROLE OF SILICON AND PHOSPHORUS (ANALOGS OF ARSENITE AND ARSENATE) IN ARSENIC PHYTOTOXICITY OF RICE

5.2.8.1 SILICON

Silicon is the second most abundant element in Earth's crust, and it acts as an analog of arsenite, and both compete with each other for their translocation through the roots of rice plants. The increased supply of silicon overcomes the expression of Lsi1 and Lsi2 transporter genes (Mawia et al., 2021; Kumarathilaka et al., 2019). The presence of a large amount of silica in the soil where rice is grown reduces the absorption of arsenic as silica inhibits the uptake of arsenic by competing with it (Mawia et al., 2021). Resistance to biotic and abiotic stress is achieved through increased silicon concentration. Both silicic acid and As(III) travel through the same route to reach the root cells of the rice plant. The aquaporin channel, Lsi1 acts as a mediator for both Si(OH)4 and arsenite absorption. When Si/As III ratio increases, arsenite uptake decreases in rice (Kumarathilaka et al., 2019).

If the amount of silica in the arsenic-contaminated rice decreases, the adsorption of arsenite increases leading to its accumulation in rice plants causing phytotoxicity (Kumarathilaka et al., 2019). The increase or decrease of arsenite absorption depends on the concentration of silica in soil because both the elements (As III and Si) compete with each other because of their chemical structure. If the concentration of arsenic in rice increases, the rate of photosynthesis is reduced and this can be overcome by supplementation of silicon (diatomaceous earth and silica gel) (Bakhat et al., 2017).

5.2.8.2 PHOSPHORUS

Soil properties such as pH, water retention capacity, soil logging, etc., amount to phosphate ions, and other ions are the important components that influence the uptake of arsenic in rice plants which is grown in an arsenic-contaminated environment (Bakhat et al., 2017). The phosphate ion is analogous to Arsenate (As V). Due to their respective chemical structures, P and As V compete to get adsorbed or absorbed by the roots of the rice plant. So, by an increase in the supply of PO_4^{3-} ion externally to the arsenic-contaminated soil where rice is grown, As V uptake by the roots is decreased and phytotoxicity due to arsenate is alleviated. Three factors influence the role of PO_4^{3-} on arsenic movement from roots to shoots of rice (Bakhat et al., 2017) (Figure 5.6).

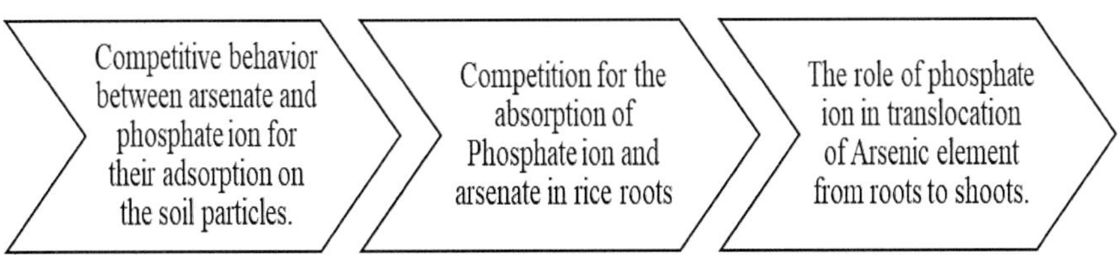

FIGURE 5.6 Role of phosphate on the translocation of arsenic in rice.

Competition between arsenic and phosphate ions in rice through the ligand exchange mechanism leads to an increment in the arsenic solubility and their increased bioavailability in the rhizosphere of rice plants. But sometimes, a decrease in the concentration of phosphate ions causes an increase in the iron plaque formation on the root site of rice which causes higher adsorption of arsenic whereas an increase in the concentration of phosphate ions results in few irons plaque formation which can increase the arsenic mobility from roots to shoots in rice plants causing phytotoxicity (Bakhat et al., 2017).

The PO_4^{3-} transporter, OsPT8 acts as a mediator for both phosphate ion and arsenate uptake through roots in rice. A higher ratio of PO_4^{3-}/As V decreases the arsenic uptake and indicates that the ratio that plays an important role in arsenic uptake in rice plants (Kumarathilaka et al., 2019).

5.2.9 ROLE OF MACRONUTRIENTS IN CONTROLLING ARSENIC ABSORPTION IN RICE (IRON AND SULFUR)

5.2.9.1 IRON (FE)

One of the essential rice mineral nutrients is iron and it helps to reduce arsenic absorption in rice plants (Mitra et al., 2017). It plays an important role in the biogeochemical process of arsenic, with the presence of iron oxyhydroxides on particulates of soil or roots of rice plants acting as strong adsorbents of arsenic (Zhao et al., 2010). The following mechanism is performed by iron to decrease arsenic availability (Figure 5.7).

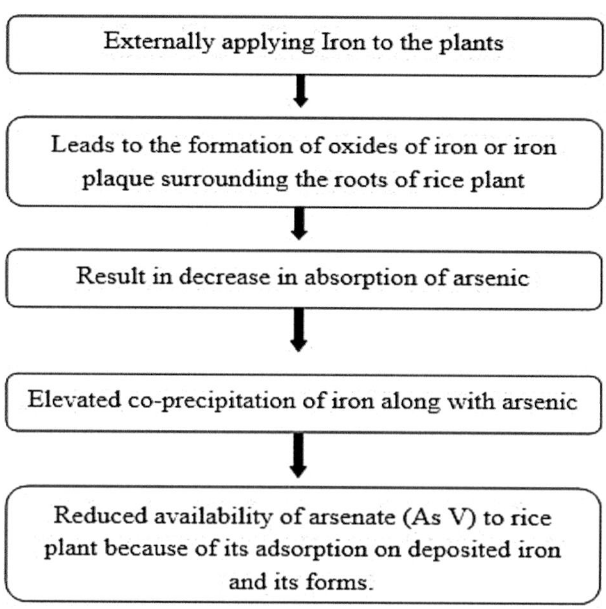

FIGURE 5.7 Mechanism of iron for reducing arsenic uptake.

Iron is a non-specific interceptor. The iron plaque has an affinity to adsorb the arsenate, hence resulting in the sequestration of arsenic, which directly decreases the availability of arsenic for translocation in rice (Zhao et al., 2010). Under redox conditions, arsenic is released back to be available for the rice plant or it is reduced to the most soluble form, i.e., As III (Arsenite) which is most toxic to rice plants (Suriyagoda et al., 2018).

5.2.9.2 SULFUR (S)

Sulfur is an important and necessary macronutrient for the growth of the rice plant. This element helps in the detoxification of arsenic through the complexation process. By forming a complex, the translocation of arsenic is reduced, and arsenic remains in the roots of rice itself (Zhao et al., 2010).

In flooded situations, sulfate present in soil and water gets reduced to S^{2-}. Arsenite (AsIII) present in rice plants combines with S^{2-} and forms As_2S_3 complex precipitation. This complex remains up to the roots of rice and thus the availability of arsenic for rice plants is decreased (Kumarathilaka et al., 2019) (Figure 5.8).

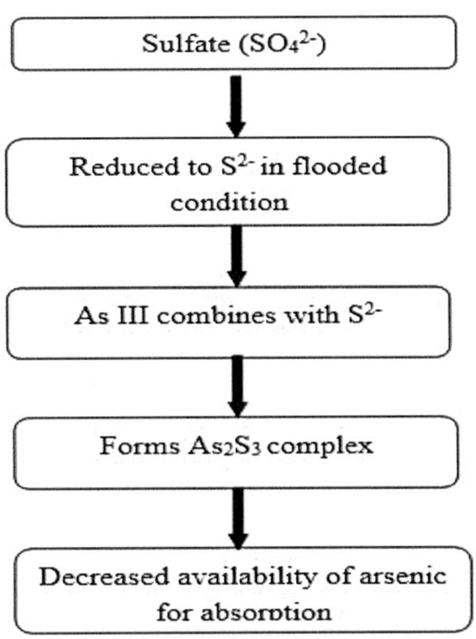

FIGURE 5.8 Mechanism of sulfur for reducing arsenic uptake in rice plants.

In addition, Sulfur changes the metabolism of the rice plant to overcome arsenic intoxication. This is possible by the addition of sulfur to the soil, which causes a reduction in the level of Lsi2 that acts as a mediator in the efflux of arsenite towards the xylem (Kumarathilaka et al., 2019). The addition of a higher amount of sulfur results in the elevation of arsenic accumulation in the roots of rice because of increased thiol ligand production and parallel increased formation of a complex in the roots of rice (Mitra et al., 2017).

5.3 DISCUSSIONS

It is so far understood that the phytotoxicity in rice plants is due to arsenic (Table 5.2), which gets bioaccumulated within the rice plant. But the several factors influence the bioaccumulation of arsenic, thus causing variation in arsenic concentration in the rice plant. Arsenic before its uptake by the plant is first interacting with the rice plant soil. Here the interaction of the soil particles with the heavy metals depends on several factors like the pH of the soil, the organic matter content of the soil, and the genotypic variation of the rice plant, which collectively determines the final concentration of arsenic in the rice plant.

TABLE 5.2 Briefing About the Impact of Arsenic on Rice Plants

Arsenic Toxicity in Rice Plants		References
Form of arsenic	Trivalent form of arsenic (As^{3+})	Bakhat et al. (2017); Rehman & Naranmandura (2012); Vasupalli et al. (2020)
Responsible factors	• Lsi1 and Lsi2 gene of rice is responsible for the absorption of arsenic. • Vacuolar transporters like OsABCC	
Source of arsenic	• Polluted reservoir, cultivated loam, and surrounding.	
Mechanism	• Arsenate reduction mechanism. • Conversion of arsenic into organic form (MMAs, DMAs, TMAs). • Binding of glutathione with reduced arsenate to form a toxic complex. • Generation of oxidative stress.	
Effects	• Height of plant is affected, reduced yield of grains, decrease in inflorescence, contaminated grains.	

The pH of the soil (both high and low pH) significantly affects arsenic uptake and bioaccumulation (Quazi et al., 2011; Chatterjee et al., 2013). When the pH is less than 5, arsenic-binding species like Fe-oxyhydroxide get soluble and enhance arsenic uptake by the rice plant (Signes-Pastor et al., 2007). Whereas high pH (near 8.5) causes an increase in negative surface charges of soil particles like OH-ions that leads to desorption of iron oxides from arsenic, which causes mobilization of arsenic towards the roots of rice, leading to arsenic accumulation in rice (Ahmed et al., 2011).

The movement of arsenic is also controlled by the organic matter and its chemical nature within the soil (William et al., 2011). Higher organic matter content reduces arsenic solubility which affects the bioavailability of arsenic in the rice plant (Paikarav et al., 2005). Another aspect of organic matter is that it has also been shown to increase arsenic mobility when the microbial activity of the soil was increased followed by a decrease in soil redox potential (Turpeinen et al., 1999).

Moreover, it has also been found that the BR11 rice variety had the highest accumulation (Reza et al., 2010) followed by IR 50, Jaya, Ratna, Lal Sanna, and Ganga-Kaveri varieties (Meharg & Rahman, 2003). This variation was due to environmental conditions, genetic differences, and the level of arsenic contamination in the waters and soils of the rice plant that was exposed (Bhattacharya et al., 2010; Zhang & Duan, 2008). Some varieties of rice with high protein content had more arsenic bio-accessibility as thiol groups strongly bind to arsenic III (Norton et al., 2009).

Various studies have been conducted by various researchers on rice from different parts of the world like the USA (Rahman et al., 2014), China (William et al., 2005), Ghana (William et al., 2005), South Asian countries (Meharg et al., 2009) for understanding the arsenic accumulation in rice plants and the factors that were influencing it.

Arsenic-contaminated rice plants are shown to trigger toxicity in humans (Table 5.3) (Mitra et al., 2017). But the human body has its mechanisms that decide the final concentration of arsenic in the body. The gastrointestinal tract (GIT) has been responsible for the biotransformation of arsenic by various mechanisms like methylation, oxidation, reduction, and thiolation (Zavala & Duxbury, 2008; Alava et al., 2015). But the acidic pH of the GIT increases the bio-accessibility of arsenic in rice (Zavala & Duxbury, 2008; Alava et al., 2015). Other key considerations that determine the arsenic concentrations are the mode (food processing, way of cooking, temperature exposure, storage conditions) by which the rice becomes available to the consumer is another interesting area of discussion (Norton et al., 2009; Mitra et al., 2017).

TABLE 5.3 Discussing the Impacts of Arsenic-Contaminated Rice on Humans

Arsenic Toxicity in Humans by Consuming Contaminated Rice		References
Form of arsenic	Majorly inorganic; As^{3+} Highly toxic form	Abir & Rodriguez (n.d.); Hong et al. (2014)
Responsible factors	The concentration of arsenic, gender, immunity of an organism, gene makeup, and nutrients.	
Mechanism of toxicity	Hinder functioning of enzymes of mitochondria and initiate Kreb's citric acid cycle and reduces the production of ATP in the cell.	
Effects	**Carcinogenic:** dermal cancer, respiratory cancer, bladder cancer, hepatic cancer, bone marrow, and lymphatic vessel cancer, prostate cancer. **Non-carcinogenic:** effect on the nervous system such as. **PNS:** insensibility, Formication of farthest end of the neuron, loss of myelin sheath. **CNS:** narcolepsy, disability of main functions like remembrance and focus. • Highly affects the white matter of the brain. • Genotoxic impacts	

5.4 CONCLUSION AND FUTURE ASPECTS

Rice comprises a major part of the daily diet in most parts of the world. Since the natural and anthropogenic activities are increased over the years, it has led to groundwater contamination posing a threat to rice farming. The environment is contaminated with significant levels of arsenic, which is causing further threats to human and animal health. Further, arsenic has been shown to cause phytotoxic effects in rice plants, which is a critical area that needs to be given attention. Scientists and policymakers should work collectively to prevent/reduce arsenic contamination in rice crops for future sustainability. Farmers and industrialists should also rethink their working strategies to minimize excessive use of groundwater and conduct industry-related activities as far away from agricultural parts as possible to reduce the threat of arsenic. Other emerging concern of arsenic phytotoxicity in rice products like rice-based infant formulations needs to be studied in depth. It is also the need of the hour to investigate the bioavailability and bioaccumulation of arsenic species in rice, as very little information is available on this aspect.

KEYWORDS

- **arsenic**
- **bioaccumulation**
- **heavy metal**
- **mechanisms**
- **phytotoxic effects**
- **phytotoxicity**
- **rice**

REFERENCES

Abir, R., & Rodriguez, V., (n.d.). *The Effects of Arsenic Exposure on the Nervous System Related Papers*. https://doi.org/10.1016/S0378-4274(03)00262-5.

Ahmed, Z. U., Panaullah, G. M., Gauch, H., McCouch, S. R., Tyagi, W., Kabir, M. S., & Duxbury, J., M. (2011). Genotype and environment effects on rice (*Oryza sativa* L.) grain arsenic concentration in Bangladesh. *Plant and Soil, 338*(1), 367–382. https://doi.org/10.1007/s11104-010-0551–7.

Alava, P., Du Laing, G., Tack, F., De Ryck, T., & Van De, W. T., (2015). Westernized diets lower arsenic gastrointestinal bioaccessibility but increase microbial arsenic speciation changes in the colon. *Chemosphere, 119*, 757–762. https://doi.org/10.1016/j.chemosphere.2014.08.010.

Awasthi, S., Chauhan, R., Srivastava, S., & Tripathi, R. D., (2017). The journey of arsenic from soil to grain in rice. *Frontiers in Plant Science, 8*, 1007. doi.org/10.3389/fpls.2017.01007.

Bakhat, H. F., Zia, Z., Fahad, S., & Abbas, S., (2017). *Arsenic Uptake, Accumulation and Toxicity in Rice Plants: Possible Remedies for its Detoxification: A Review*. https://doi.org/10.1007/s11356-017-8462-2.

Bhattacharya, P., Samal, A. C., Majumdar, J., & Santra, S. C., (2010). Arsenic contamination in rice, wheat, pulses, and vegetables: A study in an arsenic affected area of West Bengal, India. *Water, Air, and Soil Pollution, 213*(1), 3–13. https://doi.org/10.1007/s11270-010-0361-9.

Bhattacharya, P., Samal, A. C., Majumdar, J., & Santra, S. C., (2010). Uptake of arsenic in rice plant varieties cultivated with arsenic rich groundwater. *Environment Asia, 3*(2), 34–37.

Biswas, J. K., Warke, M., Datta, R., & Sarkar, D., (2020). Is arsenic in rice a major human health concern? *Current Pollution Reports, 6*(2), 37–42. https://doi.org/10.1007/s40726-020-00148-2.

Chatterjee, S., Datta, S., Mallick, P. H., Mitra, A., Veer, V., & Mukhopadhyay, S. K., (2013). Use of wetland plants in bioaccumulation of heavy metals. In: *Plant-Based Remediation Processes* (pp. 117–139). Springer, Berlin, Heidelberg. https://doi.org/10.1007/978-3-642-35564-6_7.

Chowdhury, U. K., Biswas, B. K., Chowdhury, T. R., Samanta, G., Mandal, B. K., Basu, G. C., & Chakraborti, D., (2000). Groundwater arsenic contamination in Bangladesh and West Bengal, India. *Environmental Health Perspectives, 108*(5), 393–397. https://doi.org/10.1289/ehp.00108393.

Das, H. K., Mitra, A. K., Sengupta, P. K., Hossain, A., Islam, F., & Rabbani, G. H., (2004). Arsenic concentrations in rice, vegetables, and fish in Bangladesh: A preliminary study. *Environment International, 30*(3), 383–387. https://doi.org/10.1016/j.envint.2003.09.005.

Duan, G. L., Hu, Y., Liu, W. J., Kneer, R., Zhao, F. J., & Zhu, Y. G., (2011). Evidence for a role of phytochelatins in regulating arsenic accumulation in rice grain. *Environmental and Experimental Botany, 71*(3), 416–421. doi.org/10.1016/j.envexpbot.2011.02.016.

Fitter, A. H., Wright, W. J., Williamson, L., Belshaw, M., Fairclough, J., & Meharg, A. A., (1998). The phosphorus nutrition of wild plants and the paradox of arsenate tolerance: Does leaf phosphate concentration control flowering? In: Lynch, J. P., & Deikman, J., (eds.), *Phosphorus in Plant Biology: Regulatory Roles in Molecular, Cellular, Organismic and Ecosystem Processes* (pp. 39–51). American Society of Plant Physiologists.

Gaikwad, D. J., Pal, A., & Mondal, S., (2020). Effect of arsenic toxicity on chlorophyll content and antioxidative enzymes activities in rice seedlings. *International Journal of Environment and Climate Change*, 411–421. doi: 10.9734/IJECC/2020/v10i1230318.

He, X., & Lilleskov, E., (2014). *Arsenic Uptake and Phytoremediation Potential by Arbuscular Mycorrhizal Fungi*, 259–275. https://doi.org/10.1007/978-3-662-45370-4.

Hojsak, I., Braegger, C., Bronsky, J., Campoy, C., Colomb, V., Decsi, T., & ESPGHAN Committee on Nutrition, (2015). Arsenic in rice: A cause for concern. *Journal of Pediatric Gastroenterology and Nutrition, 60*(1), 142–145.

Hong, Y., Song, K., & Chung, J., (2014). *Health Effects of Chronic Arsenic Exposure*, 245–252.

Jia, Y., Huang, H., Zhong, M., Wang, F. H., Zhang, L. M., & Zhu, Y. G., (2013). Microbial arsenic methylation in soil and rice rhizosphere. *Environmental Science and Technology, 47*(7), 3141–3148. https://doi.org/10.1021/es303649v.

Kumar, A., Singh, P. K., Srivastava, S., Dwivedi, S., Tripathi, R. D., Awasthi, G., Gupta, K., & Ansari, M. I., (2020). A comparative study on the effect of arsenic on thiolic ligands and phytochelatins in contrasting arsenic accumulating rice genotypes. *International Journal of Plant and Environment, 6*(2), 110–117. doi: 10.18811/ijpen.v6i02.03.

Kumarathilaka, P., Seneweera, S., Ok, Y. S., Meharg, A. A., & Bundschuh, J., (2020). Mitigation of arsenic accumulation in rice: An agronomical, physicochemical, and biological approach – A critical review. *Critical Reviews in Environmental Science and Technology, 50*(1), 31–71. https://doi.org/10.1080/10643389.2019.1618691.

Liu, W., Mcgrath, S. P., & Zhao, F., (2014). *Silicon has Opposite Effects on the Accumulation of Inorganic and Methylated Arsenic Species in Rice,* 423–431. https://doi.org/10.1007/s11104-013-1991-7.

Majumdar, A., & Bose, S., (2017). Toxicogenesis and metabolism of arsenic in rice and wheat plants with probable mitigation strategies. *Arsenic: Risks of Exposure, Behavior in the Environment and Toxicology*, 149–166.

Mawia, A. M., Hui, S., Zhou, L., Li, H., Tabassum, J., Lai, C., Wang, J., et al., (2021). Inorganic arsenic toxicity and alleviation strategies in rice. *Journal of Hazardous Materials, 408*, 124751. https://doi.org/10.1016/j.jhazmat.2020.124751.

Meacher, D. M., Menzel, D. B., Dillencourt, M. D., Bic, L. F., Schoof, R. A., Yost, L. J., & Farr, C. H., (2002). Estimation of multimedia inorganic arsenic intake in the US population. *Human and Ecological Risk Assessment, 8*(7), 1697–1721. https://doi.org/10.1080/20028091057565.

Meharg, A. A., & Hartley-Whitaker, J., (2002). Arsenic uptake and metabolism in arsenic resistant and nonresistant plant species. *New Phytologist, 154*(1), 29–43. doi.org/10.1046/j.1469-8137.2002.00363.x.

Meharg, A. A., & Macnair, M. R., (1991). The mechanisms of arsenate tolerance in *Deschampsia cespitosa* (L.) Beauv and *Agrostis capillaris* L. *New Phytologist, 119*, 291–297. doi.org/10.1111/j.1469-8137.1991.tb01033.x.

Meharg, A. A., & Macnair, M. R., (1992). Suppression of the high-affinity phosphate-uptake system – a mechanism of arsenate tolerance in *Holcus lanatus* L. *Journal of Experimental Botany, 43*, 519–524. doi.org/10.1093/jxb/43.4.519.

Meharg, A. A., & Rahman, M. M., (2003). Arsenic contamination of Bangladesh paddy field soils: Implications for rice contribution to arsenic consumption. *Environmental Science & Technology, 37*(2), 229–234. https://doi.org/10.1021/es0259842.

Meharg, A. A., & Zhao, F. J., (2012). *Arsenic & Rice*. Springer Science & Business Media.

Meharg, A. A., (2002). *Arsenic Uptake and Metabolism in Arsenic Resistant and Nonresistant Plant Species,* 29–43.

Meharg, A. A., Williams, P. N., Adomako, E., Lawgali, Y. Y., Deacon, C., Villada, A., Cambell, R. C., et al., (2009). Geographical variation in total and inorganic arsenic content of polished (white) rice. *Environmental Science & Technology, 43*(5), 1612–1617. https://doi.org/10.1021/es802612a.

Mitra, A., Chatterjee, S., Moogouei, R., & Gupta, D. K., (2017). *Arsenic Accumulation in Rice and Probable Mitigation Approaches: A Review*, 1–22. https://doi.org/10.3390/agronomy7040067.

Norton, G. J., Duan, G., Dasgupta, T., Islam, M. R., Lei, M., Zhu, Y., Deacon, C. M., et al., (2009). Environmental and genetic control of arsenic accumulation and speciation in rice grain: Comparing a range of common cultivars grown in contaminated sites across Bangladesh, China, and India. *Environmental Science & Technology, 43*(21), 8381–8386. https://doi.org/10.1021/es901844q.

Paikaray, S., Banerjee, S., & Mukherji, S., (2005). Sorption of arsenic onto Vindhyan shales: Role of pyrite and organic carbon. *Current Science,* 1580–1585.

Pandey, P. K., Yadav, S., Nair, S., & Bhui, A., (2002). Arsenic contamination of the environment: A new perspective from central-east India. *Environment International, 28*(4), 235–245. https://doi.org/10.1016/S0160-4120(02)00022-3.

Patel, K. S., Shrivas, K., Brandt, R., Jakubowski, N., Corns, W., & Hoffmann, P., (2005). Arsenic contamination in water, soil, sediment and rice of central India. *Environmental Geochemistry and Health, 27*(2), 131–145. https://doi.org/10.1007/s10653-005-0120-9.

Quazi, S., Datta, R., & Sarkar, D., (2011). Effects of soil types and forms of arsenical pesticide on rice growth and development. *International Journal of Environmental Science & Technology, 8*(3), 445–460. https://doi.org/10.1007/BF03326231.

Rahman, M. A., Rahman, M. M., & Naidu, R., (2014). Arsenic in rice: Sources and human health risk. In: *Wheat and Rice in Disease Prevention and Health* (pp. 365–375). Academic Press. https://doi.org/10.1016/B978-0-12-401716-0.00028-3.

Rehman, K., & Naranmandura, H., (2012). Arsenic metabolism and thioarsenicals. *Metallomics, 4*(9), 881–892. https://doi.org/10.1039/c2mt00181k.

Reza, A. S., Jean, J. S., Yang, H. J., Lee, M. K., Woodall, B., Liu, C. C., & Luo, S. D., (2010). Occurrence of arsenic in core sediments and groundwater in the Chapai-Nawabganj District, northwestern Bangladesh. *Water Research, 44*(6), 2021–2037. https://doi.org/10.1016/j.watres.2009.12.006.

Shi, T., Liu, H., Wang, J., Chen, Y., Fei, T., & Wu, G., (2014). Monitoring arsenic contamination in agricultural soils with reflectance spectroscopy of rice plants. *Environmental Science & Technology, 48*(11), 6264–6272. https://doi.org/10.1021/es405361n.

Shri, M., Singh, P. K., Kidwai, M., Gautam, N., Dubey, S., Verma, G., & Chakrabarty, D., (2019). Recent advances in arsenic metabolism in plants: Current status, challenges and highlighted biotechnological intervention to reduce grain arsenic in rice. *Metallomics, 11*(3), 519–532. doi.org/10.1039/c8mt00320c.

Signes-Pastor, A., Burló, F., Mitra, K., & Carbonell-Barrachina, A. A., (2007). Arsenic biogeochemistry as affected by phosphorus fertilizer addition, redox potential and pH in a west Bengal (India) soil. *Geoderma, 137*(3, 4), 504–510. https://doi.org/10.1016/j.geoderma.2006.10.012.

Smith, A. H., Lingas, E. O., & Rahman, M., (2000). Contamination of drinking-water by arsenic in Bangladesh: A public health emergency. *Bulletin of the World Health Organization, 78*, 1093–1103.

Suriyagoda, L. D. B., Dittert, K., & Lambers, H., (2018). Arsenic in rice soils and potential agronomic mitigation strategies to reduce arsenic bioavailability: A review arsenic in rice soils and potential agronomic mitigation strategies to reduce arsenic bioavailability: A review. *Pedosphere: An International Journal, 28*(3), 363–382. https://doi.org/10.1016/S1002-0160(18)60026-8.

Tsuji, J. S., Yost, L. J., Barraj, L. M., Scrafford, C. G., & Mink, P. J., (2007). Use of background inorganic arsenic exposures to provide perspective on risk assessment results. *Regulatory Toxicology and Pharmacology, 48*(1), 59–68. https://doi.org/10.1016/j.yrtph.2007.01.004.

Turpeinen, R., Pantsar-Kallio, M., Häggblom, M., & Kairesalo, T., (1999). Influence of microbes on the mobilization, toxicity and biomethylation of arsenic in soil.

Science of the Total Environment, 236(1–3), 173–180. https://doi.org/10.1016/S0048-9697(99)00269-7.

Upadhyay, M. K., Majumdar, A., Suresh, K. J., & Srivastava, S., (2020). Arsenic in rice agro-ecosystem: Solutions for safe and sustainable rice production. *Frontiers in Sustainable Food Systems, 4,* 53. doi.org/10.3389/fsufs.2020.00053.

Vasupalli, N., Koramutla, M. K., Aminedi, R., Kumar, V., Borah, P., Negi, M., & Deshmukh, R., (2020). Omics approaches and biotechnological perspectives of arsenic stress and detoxification in plants. *Metalloids in Plants: Advances and Future Prospects*, 249–273. https://doi.org/10.1002/9781119487210.ch13.

Williams, P. N., Price, A. H., Raab, A., Hossain, S. A., Feldmann, J., & Meharg, A. A., (2005). Variation in arsenic speciation and concentration in paddy rice related to dietary exposure. *Environmental Science & Technology, 39*(15), 5531–5540. https://doi.org/10.1021/es0502324.

Williams, P. N., Zhang, H., Davison, W., Meharg, A. A., Hossain, M., Norton, G. J., Brammer, H., & Islam, M. R., (2011). Organic matter solid phase interactions are critical for predicting arsenic release and plant uptake in Bangladesh paddy soils. *Environmental Science & Technology;45*(14), 6080–6087. https://doi.org/10.1021/es2003765.

Wright, W., Fitter, A., & Meharg, A. A., (2000). Reproductive biomass in *Holcus lanatus* clones that differ in their phosphate uptake kinetics and mycorrhizal colonization. *New Phytologist, 146*, 493–501. doi.org/10.1046/j.1469-8137.2000.00652.x.

Zagorchev, L., Seal, C. E., Kranner, I., & Odjakova, M., (2013). *A Central Role for Thiols in Plant Tolerance to Abiotic Stress*, 7405–7432. https://doi.org/10.3390/ijms14047405.

Zagorchev, L., Seal, C. E., Kranner, I., & Odjakova, M., (2013). A central role for thiols in plant tolerance to abiotic stress. *International Journal of Molecular Sciences, 14*(4), 7405–7432. doi.org/10.3390/ijms14047405.

Zavala, Y. J., & Duxbury, J. M., (2008). Arsenic in rice: I. Estimating normal levels of total arsenic in rice grain. *Environmental Science & Technology, 42*(10), 3856–3860. https://doi.org/10.1021/es702747y.

Zhang, J., & Duan, G. L., (2008). Genotypic difference in arsenic and cadmium accumulation by rice seedlings grown in hydroponics. *Journal of Plant Nutrition, 31*(12), 2168–2182. https://doi.org/10.1080/01904160802463130.

Zhao, F. J., Ma, J. F., Meharg, A. A., & McGrath, S. P., (2009). Arsenic uptake and metabolism in plants. *New Phytologist, 181*(4), 777–794. https://doi.org/10.1111/j.1469-8137.2008.02716.x.

Zhao, F., Mcgrath, S. P., & Meharg, A. A., (2010). *Arsenic as a Food Chain Contaminant: Mechanisms of Plant Uptake and Metabolism and Mitigation Strategies.* https://doi.org/10.1146/annurev-arplant-042809-112152.

Zhao, F. J., Zhu, Y. G., & Meharg, A. A., (2013). Methylated arsenic species in rice: Geographical variation, origin, and uptake mechanisms. *Environmental Science and Technology*, *47*(9), 3957–3966. https://doi.org/10.1021/es304295n.

CHAPTER 6

Translocation and Accumulation of Arsenic in Rice Grains: Role of Transporters in Tolerance and Its Incorporation in the Food Chain

JYOTIRMAY KALITA,[1] AMIT KUMAR PRADHAN,[2]
SABNOOR YEASRIN JYOTI,[3] BIKASH KALITA,[3] SAUROV JYOTI ROY,[3]
INDRAJIT KALITA,[3] MAHESH DAS,[3] HUNMILY TERONPI,[1]
PREETOM REGON,[3] and BHABEN TANTI[3]

[1]*Department of Botany, North Gauhati College, India*

[2]*Department of Botany, Pragjyotish College, India*

[3]*Department of Botany, Gauhati University, India*

ABSTRACT

Arsenic (As), a widespread hazardous metalloid enters rice plants through several transport routes that are present inside plants for the intake and translocation of other important minerals. Inside the plants, As induces toxic reactions that hinder various physiological processes, bringing down plant growth and grain yield. Moreover, the entry of As through the various transport channels into rice grains also marks its entry into the human food chain and is an undesired phenomenon because of its wide range of toxic reactions in humans, including cancer. The existing mechanism of As detoxification, immobilization, and efflux in rice plants results in a decrease in As accumulation and influx into the grains, as well as a decrease in toxic effects. Additionally, the uptake, influx, and deposition

Arsenic in Rice: A Global Menace. Garima Awasthi, Sudhakar Shrivastava, and Mahipal Singh Sankhla, (Eds.)

of As inside the rice plant being a genotypically controlled character, there are prospects of exploring rice germplasm worldwide to generate rice varieties with decreased As accumulations for cultivation in As-contaminated rice-growing regions of the world.

6.1 INTRODUCTION

Arsenic (As) is a widespread hazardous metalloid found in the crust of the earth in variable forms of minerals. It mostly occurs in a combined form with sulfur such as realgar, arsenopyrite, etc. (Smedley & Kinniburgh, 2002). The presence of As in soil sediments is assessed by the composition of the parent rock. As from the soil is mobilized by different natural and human processes, resulting in its dissolution in groundwater and increased bioavailability. Natural factors involved in As mobilization include weathering of rocks, the activity of microbes, etc., while anthropogenic activities involved in the process are extraction of groundwater, mining, etc. Other human actions, like the application of As-containing paints, herbicides, etc., contribute to the rise in As bioavailability in soil (Kalita & Tanti, 2020; Kalita et al., 2018).

Much of the bioavailable As in soils is found in inorganic forms. Arsenate [As(V)] and arsenite [As(III)] are found as the major types of As, based on the physicochemical properties of soil (Mandal & Suzuki, 2002). The concentration of As(V) is higher under oxidizing environments, whereas As(III) concentrations are higher under reducing ones. Waterlogging conditions through the majority of the time during paddy cultivations provide a reduced environment in making As(III) the ubiquitous species of As. Inorganic species of As also undergo microbial methylation in the soil leading to the emergence of organic Monomethylarsonic acid (MMA) and dimethylarsinic acid (DMA) As species.

As is known to accumulate in various plant organs from roots to shoots to grains and other edible parts (Rahman & Hasan, 2007). Plants uptake As from the soil via the roots and are then transported to different regions of the plant body. As induces toxic reactions in plants in the form of deterioration of growth and oxidative injury to tissues. In rice plants, due to the uptake of As from soil, alterations in growth and aspects of morphology as well as physiological processes take place. A decrease in chlorophyll concentration owing to As absorption results in diminished photosynthetic capacity as well as stunted root and above-ground development in rice. The incorporation of As in rice plants leads to a subsequent drop in yield (Pradhan et al., 2021; Halim et al., 2014). The amount of As in soil increases with the increase in the application of As-contaminated water through irrigation has led to a progressive decrease in grain yield year after year (Panaullah et al., 2009). Toxicity in rice by As manifests in the form of a physiological disease called 'Straighthead' that results in reduced tillering, sterile florets/spikelets, reduction in panicle length and panicle formation, and reduction in grain yield (Rahman et al., 2008). Some plants however show tolerance to As based on its accumulation in high amounts in different parts of the plant body. *Pteris vittata*, a fern, and *Sarcosphaera coronaria,* a yeast are well-known hyperaccumulators of As (Kalita et al., 2018).

Apart from its phytotoxicity, the accumulation of arsenic in edible plant tissues also leads to its incorporation into the human and animal food chains, concomitantly relaying the toxic effects (Williams et al., 2005). Although the proportion of DMA was greater at concentrations over 100 g/kg, research indicated a larger worldwide average of the very toxic inorganic forms of As present in rice grains compared to the organic forms (Carey et al., 2020; Pradhan et al., 2020). Although drinking contaminated water has been considered the dominant route of entry of As into the human body, it has been recently observed that consumption of food cultivated in As-contaminated soil contributes to the majority of the As intake, even when individuals are drinking water contaminated with As at levels higher than the WHO's acceptable limit of 10 μg/L (Mondal et al., 2021). In humans As has been known to affect the functions of various organ systems such as the integumentary system, the nervous system, the respiratory and the cardiovascular systems, the endocrine system, the hepatic system, the renal system, and the reproductive system. As also affects the immune system in humans. It is known to bring epigenetic and mutational changes in humans and is one of the causative agents of cancer (Mohammed Abdul et al., 2015).

Among the different crop plants, rice a water-loving crop plant appears to be the most vulnerable crop plant to As based on the influx and deposition of As. Various studies in the rice-cultivated areas of Asia and many other parts of the world reported increased deposition of As in the groundwater. Rice has been reported to be most susceptible to the absorption of As in comparison to that of other major staple crops due to its submerged method of cultivation (Awasthi et al., 2017). Along with the absorption, the rate of As uptake is also found to increase in rice than that of other major staple crops (Kalita et al., 2018). In South East Asian countries where the majority of the population is primarily rice consuming, the uptake route of As through the consumption of rice becomes a major threat to public health. Moreover, the loss of productivity of rice due to As toxicity also presents an impending concern to food security in undernourished regions. Hence, As appears to be a hazardous threat to the quality and quantity of rice grains.

The deposition of As inside the rice grain is not done in a single step, rather, it is the resultant effect of various stages of forward translocation by transporters and disruption in the process of translocation. The genotype of the rice plant has an impact on the deposition of As in the tissues (Norton et al., 2010). The amount of As accumulated in the grain is quite low in comparison to the overall As content absorbed by the roots (Abedin et al., 2002). Different transporters are associated at various stages starting from the influx of As through roots to its translocation into various plant tissues including the grains. Transporters vary depending upon the chemical species of arsenic being transported and the site of the transport. Certain transporters are also involved in the process of immobilization of As within the plant cells by transporting it to vacuoles to be sequestrated there by complexation with sulfur-containing peptides and preventing further translocation (Chen et al., 2017). Many transporters perform a very crucial role in the process of tolerance against As in plants, as well as in determining the concentration of metalloids in various tissues, including edible regions (Xie et al., 2019) (Figure 6.1).

6.2 TRANSPORTERS OF THE DIFFERENT FORMS OF ARSENIC

All essential elements are acquired from the soil surface through specific transporter following variable-type pathways in plants. The transporter in the process of selection and transport of element, sometimes fail to identify homologs or non-essential elements. This phenomenon results in the intake of different toxic substances like arsenic (Yang et al., 2012). In rice, no naturally occurring As transporters have been reported (Pandey et al., 2015). Arsenic in its variable forms shows the advantage of being chemically similar to some mineral elements which are readily absorbed by

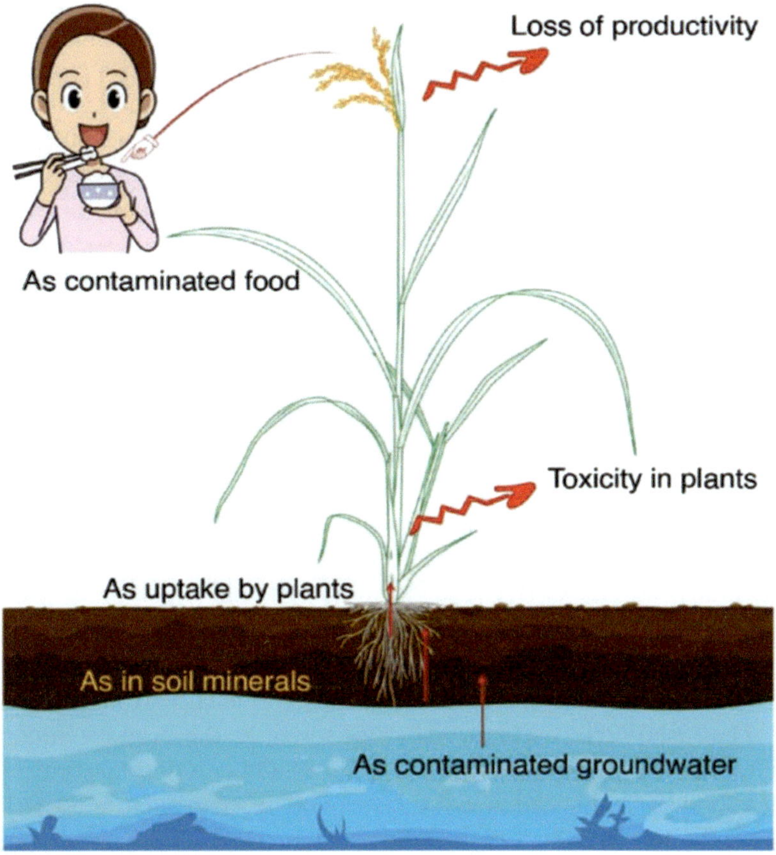

FIGURE 6.1 Effect of arsenic in plants from an environmental source.

plants (Wenzel, 2013). As(III) possesses similarity in the physiochemical features of silicon (Si), contests the Si-uptake pathway. Similarly, As(V) is analogous to phosphorus, a necessary nutrient element for the development of crop plants. Hence, using the P absorption pathways As(V) can enter the roots, from where it can be further translocated towards the xylem and various parts of the rice plant (Yang et al., 2012; Zhao et al., 2009). The influx and movement of variable As species to the subcellular compartments include the involvement of transporters such as phosphate transporters (Phts) and aquaglyceroporins which mainly includes nodulin-like intrinsic protein (NIP) and some ATP binding cassette (ABC) transporters

(Bhattacharjee et al., 2008; Garbinski et al., 2019; Mosa et al., 2012; Song et al., 2010; Zhao et al., 2009). In addition to the transporters, transcription factors like WRKY genes facilitate the uptake of As (Batista et al., 2014; Castrillo et al., 2013).

Because of the oxygenated anion structure of As(V), which is physically identical to phosphate ions, it influxes into the plant roots primarily through phosphate transporters. Different plant species have been found to contain Pi transporters with a high preference for As(V). Two Pht1 isoforms, AtPht1;1 (Pi transporter 1;1) and AtPht1;4, have been identified in *Arabidopsis thaliana* for As uptake (V). The OsPht1;1 and OsPht1;8 have been identified in the roots of *Oryza sativa*, and they have a high preference for As(V) uptake (Ma et al., 2008). Overexpression of OsPht1;8 was reported to induce import and efflux of As(V) into the root and shoot of *O. sativa* (Jia et al., 2011; Wu et al., 2011).

Although As(V) is abundant in aerobic soils, it is converted to As(III) in anaerobic circumstances such as those found during submerged rice cultivation. After being transported to the plant cell, arsenic is typically found in the form of As(III). Aquaporins (AQP), plant membrane proteins, have been identified for the transport of As(III) into the plant tissue (Li et al., 2014). Aquaglyceroporins provide a high affinity to As(III) and provide a channel for the transport to plant cells, especially from submerged soil. The OsNIP2;1 (Lsi1), a NIP transporter, is the primary aquaporin implicated in As(III) uptake (Ma et al., 2008). Other NIP transporters have been found for enhanced As(III) absorption in rice, such as OsNIP1;1, OsNIP2;2 (OsLsi6), OsNIP3;1, OsNIP3;2, and OsNIP3;3. Other AQPs for As(III) absorption and transport in rice have been found, including the plasma membrane intrinsic protein (PIP) transporter OsPIP2;4, OsPIP2;6, and OsPIP2;7. The uptake of As has also been described using the Natural resistance-associated macrophage protein 1 (NRAMP1) transporter, a membrane channel protein implicated in metalloid transport and uptake (III).

Different methylated arsenic forms, primarily MMA and DMA, have been discovered to be prominent in agricultural soils in addition to As(V) and As(III) (Li et al., 2009). Increased usage of pesticides and insecticides with the involvement of microbes has been reported to be the major cause of high As methylation (Awasthi et al., 2017). Foodgrains have recently been revealed to have higher levels of methylated As than inorganic As (Kalita et al., 2018). The accumulation and translocation of MMA and

DMA are also mediated by NIP transporters. The absorption of methylated As in rice plants is primarily mediated by Lsi1 (Li et al., 2009). Because MMA and DMA have a structural similarity to glycerol, Lsi1, an aqua-glyceroporin, provides a pathway for their uptake into the plant cell.

6.3 TRANSLOCATION OF ARSENIC IN RICE PLANTS – FROM ROOTS TO GRAINS

The translocation of As into the plant cell is normally regulated by its quantity, complexation ability, and sequestration, including the rate of flow through the xylem tissues. The concentration of As in rice generally decreases in the manner of its translocation from roots to the grains, i.e., root>shoot>grains (Suriyagoda et al., 2018). The absorption and transfer of various As forms to the roots and shoot and then to the grain vary by genotype and also rely on the development of plants (Bhattacharya et al., 2010).

The majority of rice genotypes retain or possess a significant amount of harmful form of As in the roots. However, only a little quantity of As is translocated and accumulates in the rice plant's shoots and grains (Pandey et al., 2015; Williams et al., 2007). A different study reported that As is accumulated more effectively or in higher amounts in the shoots and grain than in other cereals. Generally, the organic forms of As like DMA are transported more conveniently to the above-ground portion of rice plants than that of its inorganic forms (Raab et al., 2007). The dominating forms of As in rice grains are As(III) and DMA. Mostly, DMA is accumulated or distributed towards the outer part of rice grain till the endosperm layer while As(III) is specifically accumulated specifically towards the ovular region of vascular tissue (Carey et al., 2010). The variation in localization, movement, and deposition of different forms of As is similar in the majority of the plant species (both monocot and dicot). Thereby, this process further suggests an identical basic transport mechanism to various forms of As in different plants (Raab et al., 2007).

The translocation of As and its deposition in the grains part of the rice has been majorly the outcome of the forward transport of As in various stages, i.e., from soil to shoots via roots and finally into the rice grains. The translocation and accumulation are mediated by a number of transporters, and the inherent mechanisms in the plant that immobilize the metalloid in various parts of the plant body or extrude it back to the soil, thereby reducing its forward movement. The following segments of this chapter discuss the various stages through which As is translocated in the plant body, followed by a discussion of the mechanisms that prevent its forward translocation at various stages during its movement.

6.3.1 FORWARD TRANSPORT OF ARSENIC

6.3.1.1 ARSENIC UPTAKE BY THE ROOT CELLS

The uptake of different forms of As into the root cells occurs through different mechanisms. As(V), in general, enters into rice roots mainly through PHTs such as through phosphate transporter1 (PHT1) in common (Mishra et al., 2017; Murugaiyan et al., 2021). As per the reports, 13 PHTs of the PHT1 family are found to be associated with the absorption and transport of inorganic phosphate (Pi) in rice (Ye et al., 2017). As(III) the other bioavailable form of As utilizes one of the major membrane channel transporter proteins, the NIPs aquaporin channels (Mishra et al., 2017; Mitani et al., 2008). Three major groups of NIPs (NIP1, NIP2, and NIP3) that possess similar forms of selectivity and consistency of substrate and composition of amino acid are mostly found associated with the absorption of As(III). OsNIP2;1, found towards the dorsal part of inner plasma membranes in the outer and inner cells of rice majorly facilitates the influx of As(III) into the root tissues of rice plants (Bienert et al., 2008).

The metabolic actions of soil microorganisms are also associated with the conversion of As(V) to As(III) which subsequently gets transformed into MMA and DMA. The silicic acid transporter, Lsi1 is associated with the influx process of DMA and MMA into the crops (Suriyagoda et al., 2018). However, the level of absorption of DMA and MMA by the plant roots is lesser than those of inorganic species of As, which further gets lowered with the higher number of methyl groups (Zhao et al., 2013). Although in the rice roots, the pentavalent form of MMA is reduced to trivalent MMA to some extent, only the pentavalent form is transported to shoots (Guillod-Magnin et al., 2018). Likewise, the pentavalent form of DMA is moveable in plants and can be easily translocated to the shoot from rice roots.

6.3.1.2 TRANSLOCATION OF AS FROM ROOT TO SHOOT

Rice has a translocation factor of approximately 0.8, which is the highest in comparison to that of other cereals such as barley (0.2), and wheat (0.1) (Kalita et al., 2018; Ma et al., 2008). Aquaglyceroporin channels are mainly associated with the transfer of As(III) from roots to shoots, but the assimilation of As(III) from the soil is regulated by Lsi1, Lsi2 (OsNIP3;2) which plays a significant role in the translocation of As(III) to shoots. Lsi2 present towards the proximal sides of endodermal and exodermal root cells directs the flow of arsenite to the xylem (Ye et al., 2017; Zhao et al., 2010). Lsi1 and Lsi2 regulate the transfer of As(III) from vascular xylem vessels, finally to shoots mainly based on the physiological mechanism of transpiration pull that occurs within the parenchymatous cells of xylem tissues (Murugaiyan et al., 2021).

The absorption and translocation of As(V) through the xylem tissues, increases or gets enhanced solely as per the constitutive expression pattern of OsPHT1;8 (Wu et al., 2011). After absorption into the roots, As(V) gets reduced to As(III), which then subsequently reaches the xylem via the As(III) effluxer (silicic acid). As(V) reductase enzymes, High As Content 1 (HAC1) plays an essential role in reducing As(V) to As(III) inside rice roots (Yanshan Chen et al., 2017). This enzyme mainly functions towards the epidermal layer of the root cell (Chao et al., 2014) and the endodermal layer adjacent to the pericycle. Previous research has revealed that arsenite is the major As element collected in rice xylem sap, regardless of whether it is translocated in the form of As(V)/As(III) (Carey et al., 2010; Su et al., 2010; Zhao et al., 2009; Ma et al., 2008).

The methylated form of As (DMA) is mostly translocated across the vascular tissues in rice, possibly due to the interaction of OsPTR7. *OsPTR7* is found to express a higher pattern during arsenic uptake, in shoots than that of roots at the initial stage. During the developmental stage of rice plants, it also showed expression in the leaf blade or leaf sheath, besides increased expression in roots. Genetic modification of OsPTR7 led to the decreased deposition of DMA in shoots and grains, which thereby is considered to be a long-distance transporter of DMA in rice crops (Tang et al., 2017). It has also been reported that the mobility of DMA increases as it gets translocated from roots to higher parts of the plant in the shoots (Murugaiyan et al., 2021).

6.3.1.3 AS TRANSLOCATION TO GRAINS

The translocation of different As species to the grains is relatively less worked out. The vascular bundles are responsible for the movement of arsenic along with other mineral elements to the grains. In the As hyperaccumulator *Pteris vittata*, a large amount of As was detected in the xylem tissue during an experiment where the plants were grown using polluted soil from the mining area (Chen et al., 2005). This indicates the function of the xylem in the forward translocation of As. *P. vittata* is however known to accumulate As in its fronds. The speciation of As was also not determined in the study.

The phloem is also vital for the transmission of arsenic from vegetative tissue to grain. Of the two As species mostly reported in grains, i.e., As(III) and DMA, phloem transport was reported to account for 90% of the former and 55% of the latter which suggests that DMA is deposited to the grains via both xylem and phloem mediated pathways, whereas for As(III) into the grains, phloem serves to be the sole pathway (Carey et al., 2010). The sap of phloem has a high pH of about 8, which prevents the thiol-complexation of As(III), unlike in the xylem sap (Awasthi et al., 2017). Hence As(III) is present in a free state in the phloem sap and is readily translocated through this pathway into the grains. As(III) transport into the phloem has been determined to be mediated by the inositol transporters *AtINT2* and *AtINT4* in *Arabidopsis*. These transporters thereby regulate the deposition of As(III) in grains (Duan et al., 2015). Possibilities of identical transporters mediating As(III) transfer into phloem also exist in rice. In addition, the OsPTR7 transporter associated with the transport of DMA to far distances is assumed to have a role in its deposition into the rice grains (Tang et al., 2017).

6.3.2 DISRUPTION OF FORWARD TRANSPORT OF AS

6.3.2.1 ROLE OF THIOL COMPLEXATION BY PHYTOCHELATINS, GLUTATHIONE, AND VACUOLE SEQUESTRATION IN DISRUPTING FORWARD MOVEMENT OF AS

Sulfur is required for the influx of As from soil to roots and transportation to various tissues including the rice grain. It has been observed that sulfur helps to promote the development of Fe-plaque on roots (Hu et al., 2007). Rice plants have a natural predisposition to produce Fe-plaques on their root surfaces as a result of radial oxygen loss (ROL). During the absorption of As from the soil into the roots, the iron plaques on the root surface serve as an obstruction. Enhancing the development of Fe-plaque by sulfur increases the impermeability of the root surface to As(V).

Although As(V) is the most common type of As, it is immobilized by Fe-plaque formation in roots (Awasthi et al., 2017). However, As(III) is one of the most readily available chemical species during submerged paddy farming. Despite the fact that a portion of the As(III) in the soil gets oxidized to As(V) as a result of ROL's creation of an oxidizing environment in the rhizosphere, a considerable amount of As(III) finds its way into the rice roots (Punshon et al., 2017). Moreover, rice can reduce the As(V) that enters the plant to As(III) by the activity of arsenate reductases (ARs) present in its roots (Shi et al., 2016). As a result, As(III) accounts for the majority of the inorganic As transferred to the rice plant's aboveground tissues.

Rice plants use thiol-complexation mechanisms to prevent As(III) from being translocated, resulting in its immobilization. Plants have a class of thiol-containing metal-binding peptides called phytochelatins (PC) that play a key part in this process. The enzyme phytochelatin synthase (-glutamylcysteine dipeptidyl transpeptidase) produces PCs by transferring the -glutamylcysteine dipeptide moiety of glutathione to another glutathione that acts as an acceptor or a developing [Glu(-Cys)]n-Gly chain. During the manufacture of PCs, metal ions are required to activate phytochelatin synthase. The PCs then bind those activator metals, thus stopping the enzyme reaction (Grill et al., 1989). PCs contain the common formula (Glu-Cys)n-Gly, with n extending from 2–11, and are referred to as PC_2, PC_3, and so on (Hartley-Whitaker et al., 2001). As(III) is chelated by PCs to form As-PC complexes whereby toxic effects of the metalloid are prevented. These As-PC complexes are then carried to vacuoles, where they are sequestered. ABCC-transporters, which are found in dissimilar plant species, are associated with the transfer of these complexes to vacuoles (Song et al., 2010).

In a number of plant species, PC formation has been found to increase in response to As absorption. The difference in the dominant type of PC has been observed between As-tolerant and susceptible plants with PC_2 being the dominant form in tolerant plants whereas PC_3 is the dominant form in susceptible plants (Kalita et al., 2018). Production of PCs in response to As uptake has also been reported in rice. The concentration of PC in the roots was shown to be related to the concentration of inorganic As (Batista et al., 2014).

The enzymes glutathione S-transferases (GST), also a precursor in PC synthesis, comprise a diverse group of isozymes that are involved in catalyzing the conjugation of xenobiotics with glutathione (GSH) leading to their detoxification (Bianucci et al., 2017). Stimulation of GST during As stress was also observed in *Luffa* seedlings (Singh et al., 2015). Detoxification mechanisms through conjugation with GSH are well known (Kumar & Trivedi, 2018). During As uptake, conjugation of As with GSH leading to its detoxification has been identified in *Eichhornia crassipes* (de Souza Reis et al., 2020). Like the complexes with PCs, As-GSH conjugates have also been reported to be sequestrated in vacuoles (Zhang et al., 2018). The Lambda-class of GSTs is found to be connected in response to As stress in different species of plants, of which rice contains three members belonging to the family of Lambda gene, viz. the *OsGSTL1*, *OsGSTL2*, and *OsGSTL3* (Gautam et al., 2020). Thus, sulfur or sulfur-containing thiols have been found to serve an important function in the uptake regulation, assimilation, and sequestration of As, thereby reducing its translocation from different tissues to the rice grains.

Although it is an extensively considered fact that As(V) is reduced to As(III) in the cells of plant roots, the predominance of As(V) was found in the xylem region of a cross-sectioned root (Seyfferth et al., 2011). The retention of As(V) in high amounts in the straw of rice grown in an arsenate solution for several days (Abedin et al., 2002). This study further highlights the importance of further analysis related to the reduction of As(V) to As(III) in root cells.

6.3.2.2 ROOT EXUDATION

Apart from sequestration in vacuoles, As(III) has another destination in plants. The ability of certain plants to efflux As(III) back into the growing medium has been demonstrated. The silicon transporter Lsi1, which is also engaged in As(III) absorption and movement in the rice plant, has been discovered as a transporter implicated in the transport of As(III) to the external part of rice. However, another silicon transporter that is engaged in

As(III) transport within the rice plant is not involved in As(III) efflux to the growth media. As a result, Lsi1 is a bidirectional As(III) transporter (Zhao et al., 2010). In rice and tomato plants, rapid reduction of As(V) to As(III) and efflux of the latter into the growth media have been reported. The rate of efflux however declined with time, suggesting that efflux to the growth medium is an initial response to As uptake, which is later reduced due to the sequestration of As(III) to the vacuoles by complexation with thiols (Xu et al., 2007). Furthermore, the action of HAC (High Arsenic Content) reductases has been reported to be linked to As(III) efflux in plants, as well as the reduction of As(V) to As(III). HAC1, which gets deposited in the root epidermis and pericycle of *A. thaliana*, converts As(V) to As(III), which is then translocated back into the external media (Chao et al., 2014). Rice roots include the OsHAC1;1, OsHAC1;2, and OsHAC4 reductases, which decrease As(V) and cause As(III) to be extruded into the external media (Shi et al., 2016; Xu et al., 2017). Mutants of these reductases exhibit lower As(III) efflux into the growth medium and enhanced As(III) deposition in shoots and other aboveground tissues. About 15–20% of the overall efflux is carried out by the Lsi1 transporter, which was shown to transport As(III) out of the roots (Zhao et al., 2010). As a result, alternative transporters implicated in As(III) efflux in rice have yet to be discovered.

6.3.2.3 ROLE OF NODES

Plant stems are characterized by the presence of nodes and internodes. Nodes are important as they are the main point of leaf attachment and flowering. It also acts as a regulator in the process of nutrient regulation and circulation in plants (Ma & Yamaji, 2015). The vascular system, which is important for the distribution of components in plants, is also found in nodes. The vascular tissues are also located in nodes, essential for the distribution of elements in plants. The storage and delivery of As to rice grains are similarly regulated by nodes (Ma & Yamaji, 2015; Zhao et al., 2014). Comparatively, experimental evidence has suggested a higher accumulation of As in the node than in the internode and leaves of plants. Synchrotron μXray fluorescence (μ-XRF) technique has elucidated the role of Lsi2 mutant and phytochelatins (PC) in As distribution. In rice, the node also acts as an As(III) filter, thus minimizing the As(III) supply into the grain (Chen et al., 2015; Moore et al., 2014). In addition, the

ABCC transporter, found within the nodes of phloem cells, i.e., inside the tonoplast, regulates the trafficking of PC-As(III) (phytochelatin-arsenite complex) into vacuoles (Moore et al., 2014; Song et al., 2014). Vacuolar PC–As(III) transporter OsABCC1 was reported to be localized in nodes and inhibit the transport of As inside the grains by transferring the PC-As complex inside the vacuoles of phloem cells. In rice, accumulation of As was also reported in the phloem companion cells of nodes, suggesting further translocation and restriction of As into the grains and flag leaves (Murugaiyan et al., 2021) (Figure 6.2).

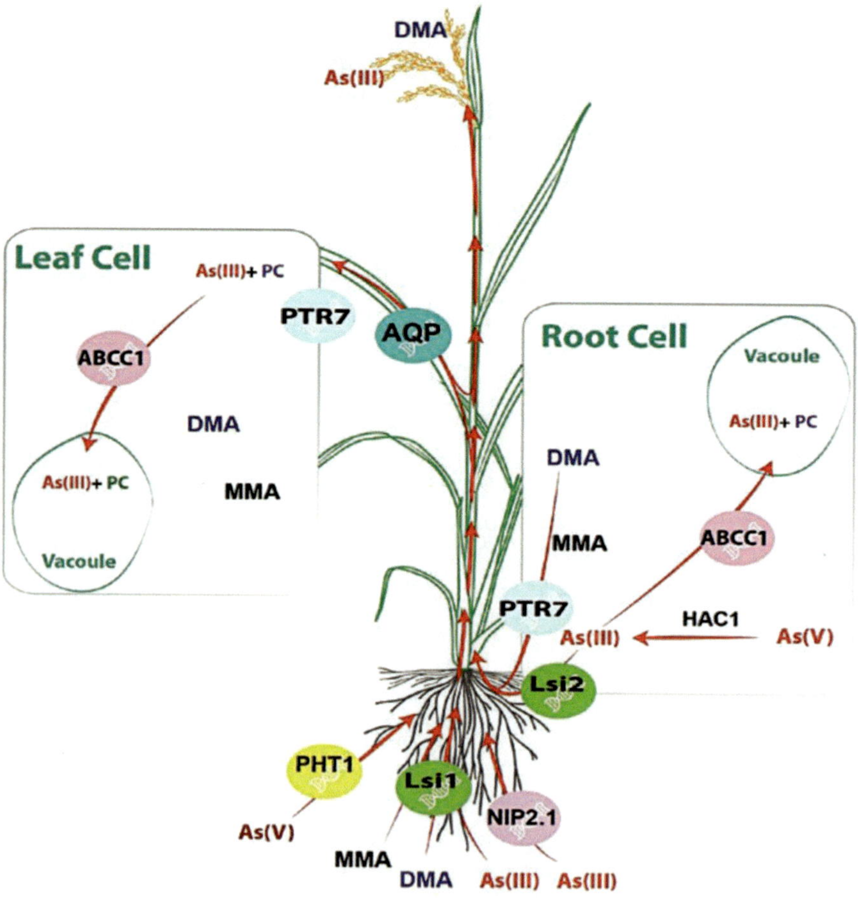

FIGURE 6.2 Transport of different forms of arsenic from soil to rice grains.

6.4 GENETIC REGULATION OF TRANSPORT

The transport of different forms of As into rice plants is facilitated by various phosphate or silicon transporters. According to the evidence of physiological studies, the involvement of AQP in As transport in rice has been established (Meharg & Jardine, 2003). Although the specific method of As movement in plants is unknown, investigations have suggested the involvement of nodulin26-like intrinsic proteins (NIPs) in As trafficking. Since, As is not a required nutrient for plants, regulation of the expression of its transporters is critical for As tolerance. The most important category of genes associated with As regulation in plants are the transcription factors (TFs) and signaling regulators which are responsible for the expression of downstream genes. The members of the MYB family that play an essential role in the management of abiotic and biotic factors also serve a crucial role in the management of As in the plants. About 16 transcription factor related to As detoxification has been reported, where the expression of MYB encoding genes Os10g0562100, Os01g0685400, and Os04g0593200 was highly induced (Nguyen et al., 2014). In addition, OsARM1, an R2-R3 MYB transcription factor was reported to be associated with the management of transporter genes involved in the transfer and detoxification of As in rice (Wang et al., 2017). OsARM1 was highly expressed under As(III) toxicity (Wang et al., 2017). In rice, the predominant expression of *OsARM1* in the phloem of upper and basal nodes was discovered by GUS staining. Further, comparative phenotypic, genetics, and biochemical studies have proven the role of *OsARM1* in the regulation of As tolerance in rice plants, by controlling the absorption and translocation of As from root to above-ground parts of the plant. Transgenic line *OsARM1-KOs*, translocated more As from root to shoots, under decreased As(III) conditions (2 μM). Whereas, in the *OsARM1-OEs* line, accumulation of As was higher in the roots in response to high As(III) (25 μM). Furthermore, when compared to the wild type, *OsLsi1, OsLsi2, and OsLsi6,* which encode important As transporters, were considerably negatively regulated in *OsARM1-OEs* and elevated in *OsARM1-KOs*.

Heavy metal stressors, such as As toxicity, enhance glutathione (GSH) levels, which facilitates positive induction and increased expression of glutathione S transferase (GST) (Shri et al., 2009). Phytochelatins (PCs) are generated non-translationally from GSH, form complexes with As, and get sequestered into the vacuoles via transporters like ABCC1/ABCC2. The basic helix-loop-helix (bHLH) transcription factors gene (*Os10g0556200*) was found to be positively induced in As(V) stressed rice (Nguyen et al., 2014). Moreover, roots exposed to As(V) showed a 40-fold increase in bZIP10 (*Os01g0859300*) expression in rice.

Influx and translocation in plants are highly influenced by As speciation. In plants, the conversion of As(V) to As(III) is regulated by AR, whereas (High As Content1) reduces As(V) to As(III) in the outer cell layer of roots (Chao et al., 2014). HAC1 orthologs *OsHAC1;1* and *OsHAC1;2* are responsible for the reduction of As(V) to As(III) and are highly expressed in roots, thus facilitating the efflux of As(III) from roots (Kalita et al., 2018; Shi et al., 2016). However, OsHAC1;1 is abundant in the epidermis, root hairs, and pericycle as well as in stems, whereas OsHAC1;2 is abundant in the epidermis, outer cortical layers, and endodermis. The OsHAC4 is also expressed in the maturation and elongation zone of the epidermis and exodermis of rice (Xu et al., 2017). The mutation of all these genes results in a decrease in As(V) reduction to As(III) in roots and as a consequence, the level of As in shoots increases. Overexpression of these genes, on the other hand, shows the opposite effect (Shi et al., 2016; Xu et al., 2017). Glutaredoxins (Grxs) proteins are involved in different cellular processes like the regulation and protection mechanisms under oxidative stress (Awasthi et al., 2017). Two Grx viz. *OsGrx_C7* and *OsGrx_C2.1* are reported to be present in rice and overexpression of *OsGrx _C7* and *OsGrx _C2.1* in *A. thaliana* imparted As tolerance and reduced As deposition in seeds and shoot tissues when compared to wild-type (Verma et al., 2016). As a result, *OsGrx_C7* and *OsGrx_C2.1* are significant determinants of plant As-stress responses (Pradhan et al., 2020; Awasthi et al., 2017).

The involvement of small, non-coding miRNAs is another regulatory mechanism reported in plants to combat oxidative stresses. Analysis of the role of *Osa-miR156j* in different growth stages (seedling, tillering, and flowering) and tissues from various parts (roots, stem, and leaves) of rice plants under As(III) stress (25 μM) revealed that the *Osa-miR156j* play a vital role in rice under As(III) stress, as *Osa-miR156j* influences metabolic processes in rice plants at various developmental stages (Pandey et al., 2020).

6.5 CONCLUSION

As contamination of agricultural soil and groundwater which has led to the uptake of the metalloid by rice plants creates two significant problems. As a toxic metalloid, it causes toxicity in rice plants, altering their growth and physiology, and resulting in a decrease in crop production (Kalita et al., 2018). Moreover, As eventually reaches the rice grains and accumulates there due to translocation by different transporters (Williams et al., 2005). As a result, As gets into the food chain, and its consumption leads to toxic effects in humans (Mohammed Abdul et al., 2015). The inorganic forms of As are more toxic than the organic forms of the element. In the rice grain, the inorganic As(III) and the organic methylated form DMA is the most readily available forms of As (Carey et al., 2010).

Rice plants have a variety of methods for dealing with As-induced abiotic stress and lowering As accumulation in the tissues. As(III) is complexed with thiol-containing peptides phytochelatins or glutathione, which is subsequently transported by ABCC transporters to vacuoles and sequestrated there. Eventually, the entry of As(III) into rice grains is inhibited by this detoxifying mechanism. The inorganic As(V) is also reduced to As(III) by reductases in the plants such as HAC1 and thereafter follows the same fate. Through the action of aquaglyceroporin Lsi1, As(III) is also effluxed into the growing medium by plant root cells, leading to the reduction of As concentration in plant tissues.

As transporters are reported to serve a significant role in As detoxification in rice plants, as well as the lowering of As concentration in grains and other aboveground tissues. Considering the importance of rice, as well as its higher vulnerability to As uptake and accumulation, it's critical to develop rice cultivars that can not only withstand As stress but also accumulate lower levels of the metalloid in their grains, especially for cultivation in areas where As contamination of soil and groundwater is widespread. A transgenic approach to developing rice accumulating lower amounts of As in grains has made use of As(III) vacuolar sequestration genes *ScYCF1* from yeast and *OsABCC1* from rice, along with a prokaryotic γ-glutamylcysteine synthetase which is involved in glutathione biosynthesis, resulting in rice that uptake 70% less As in grains without deteriorating agronomic traits (Deng et al., 2018). The HAC1 may be a crucial gene for the synthesis of low-As foods in rice (Chao et al., 2014). These findings not only indicate the importance of As(V) reduction and As(III) detoxification strategies involving thiol complexation and vacuole sequestration in reducing As content in grains, but they also point to ways to develop rice for use in As-contaminated areas. In comparison to a mutation in Lsi1, a mutation in Lsi2 was observed to result in a greater reduction of As accumulation into the grain (Ma et al., 2008).

The above finding suggests that xylem loading of As is a more vital step in determining the As deposition in rice grains as compared to As(III) influx by roots or its efflux to the growth medium. A global exploration for naturally low grain As accumulating rice led to the identification of such cultivars only in the tropical regions of the Eastern half of the Southern Hemisphere in East Africa and the Southern islands of Indonesia (Carey et al., 2020). To take advantage of the inherent variance among rice cultivars for use in breeding efforts targeted at producing acceptable variations for As contaminated rice growing locations around the world, the underlying determinants for low grain As accumulating features must be determined.

KEYWORDS

- **arsenic**
- **arsenic contamination**
- **bioaccumulation**
- **rice**
- **translocation**
- **transporters**
- **γ-glutamylcysteine**

REFERENCES

Abedin, J., Cresser, M. S., Meharg, A. A., Feldmann, J., & Cotter-Howells, J., (2002). Arsenic accumulation and metabolism in rice (*Oryza sativa* L.). *Environmental Science and Technology, 36*(5), 962–968. https://doi.org/10.1021/es0101678.

Awasthi, S., Chauhan, R., Srivastava, S., & Tripathi, R. D., (2017). The journey of arsenic from soil to grain in rice. *Frontiers in Plant Science, 8*, 1–13. https://doi.org/10.3389/fpls.2017.01007.

Batista, B. L., Nigar, M., Mestrot, A., Rocha, B. A., Júnior, F. B., Price, A. H., Raab, A., & Feldmann, J., (2014). Identification and quantification of phytochelatins in roots of rice to long-term exposure: Evidence of individual role on arsenic accumulation and translocation. *Journal of Experimental Botany, 65*(6), 1467–1479. https://doi.org/10.1093/jxb/eru018.

Bhattacharjee, H., Mukhopadhyay, R., Thiyagarajan, S., & Rosen, B. P., (2008). Aquaglyceroporins: Ancient channels for metalloids. *Journal of Biology, 7*(9). https://doi.org/10.1186/jbiol91.

Bhattacharya, P., Samal, A. C., Majumdar, J., & Santra, S. C., (2010). Accumulation of arsenic and its distribution in rice plant (*Oryza sativa* L.) in Gangetic West Bengal, India. *Paddy and Water Environment, 8*(1), 63–70. https://doi.org/10.1007/s10333-009-0180-z.

Bianucci, E., Furlan, A., Tordable, M. D. C., Hernández, L. E., Carpena-Ruiz, R. O., & Castro, S., (2017). Antioxidant responses of peanut roots exposed to realistic groundwater doses of arsenate: Identification of glutathione S-transferase as a suitable biomarker for metalloid toxicity. *Chemosphere, 181*, 551–561. https://doi.org/10.1016/j.chemosphere.2017.04.104.

Bienert, G. P., Thorsen, M., Schüssler, M. D., Nilsson, H. R., Wagner, A., Tamás, M. J., & Jahn, T. P., (2008). A subgroup of plant aquaporins facilitate the bi-directional diffusion of $As(OH)_3$ and $Sb(OH)_3$ across membranes. *BMC Biology, 6*(III), 1–15. https://doi.org/10.1186/1741-7007-6-26.

Carey, A. M., Scheckel, K. G., Lombi, E., Newville, M., Choi, Y., Norton, G. J., Charnock, J. M., et al., (2010). Grain unloading of arsenic species in rice. *Plant Physiology, 152*(1), 309–319. https://doi.org/10.1104/pp.109.146126.

Carey, M., Meharg, C., Williams, P., Marwa, E., Jiujin, X., Farias, J. G., De Silva, P. M. C. S., et al., (2020). Global sourcing of low-inorganic arsenic rice grain. *Exposure and Health, 12*(4), 711–719. https://doi.org/10.1007/s12403-019-00330-y.

Castrillo, G., Sánchez-Bermejo, E., De Lorenzo, L., Crevillén, P., Fraile-Escanciano, A., Mohan, T. C., Mouriz, A., et al., (2013). WRKY6 transcription factor restricts arsenate uptake and transposon activation in *Arabidopsis*. *Plant Cell, 25*(8), 2944–2957. https://doi.org/10.1105/tpc.113.114009.

Chao, D. Y., Chen, Y., Chen, J., Shi, S., Chen, Z., Wang, C., Danku, J. M., et al., (2014). Genome-wide association mapping identifies a new arsenate reductase enzyme critical for limiting arsenic accumulation in plants. *PLoS Biology, 12*(12). https://doi.org/10.1371/journal.pbio.1002009.

Chen, T., Huang, Z., Huang, Y., & Lei, M., (2005). Distributions of arsenic and essential elements in pinna of arsenic hyperaccumulator *Pteris vittata* L. *Science in China, Series C: Life Sciences, 48*(1), 18–24. https://doi.org/10.1360/04yc0011.

Chen, Y., Han, Y. H., Cao, Y., Zhu, Y. G., Rathinasabapathi, B., & Ma, L. Q., (2017). Arsenic transport in rice and biological solutions to reduce arsenic risk from rice. *Frontiers in Plant Science, 8*. https://doi.org/10.3389/fpls.2017.00268.

Chen, Y., Moore, K. L., Miller, A. J., McGrath, S. P., Ma, J. F., & Zhao, F. J., (2015). The role of nodes in arsenic storage and distribution in rice. *Journal of Experimental Botany, 66*(13), 3717–3724. https://doi.org/10.1093/jxb/erv164.

De Souza, R. I. N. R., Alves De, O. J., Ventrella, M. C., Otoni, W. C., Marinato, C. S., & Paiva De, M. L., (2020). Involvement of glutathione metabolism in *Eichhornia crassipes* tolerance to arsenic. *Plant Biology, 22*(2), 346–350. https://doi.org/10.1111/plb.12988.

Duan, G. L., Hu, Y., Schneider, S., McDermott, J., Chen, J., Sauer, N., Rosen, B. P., Daus, B., Liu, Z., & Zhu, Y. G., (2015). Inositol transporters AtINT2 and AtINT4 regulate arsenic accumulation in *Arabidopsis* seeds. *Nature Plants, 1*(1), 1–6. https://doi.org/10.1038/nplants.2015.202.

Garbinski, L. D., Rosen, B. P., & Chen, J., (2019). Pathways of arsenic uptake and efflux. *Environment International, 126*, 585–597. https://doi.org/10.1016/j.envint.2019.02.058.

Gautam, A., Pandey, A. K., & Dubey, R. S., (2020). Unravelling molecular mechanisms for enhancing arsenic tolerance in plants: A review. In: *Plant Gene* (Vol. 23). Elsevier B.V. https://doi.org/10.1016/j.plgene.2020.100240.

Grill, E., Löffler, S., Winnacker, E. L., & Zenk, M. H., (1989). Phytochelatins, the heavy-metal-binding peptides of plants, are synthesized from glutathione by a specific gamma-glutamylcysteine dipeptidyl transpeptidase (phytochelatin synthase). *Proceedings of the National Academy of Sciences of the United States of America, 86*(18), 6838–6842.

Guillod-Magnin, R., Brüschweiler, B. J., Aubert, R., & Haldimann, M., (2018). Arsenic species in rice and rice-based products consumed by toddlers in Switzerland. *Food Additives and Contaminants – Part A Chemistry, Analysis, Control, Exposure and Risk Assessment, 35*(6), 1164–1178. https://doi.org/10.1080/19440049.2018.1440641.

Halim, M., Niger, M., Ghosh, M., Akhter, N., & Hossain, F., (2014). Interaction effects of arsenic and phosphorus on seedling growth, leaf pigments and leaf protein of rice cultivars under hydroponic culture. *Journal of the Asiatic Society of Bangladesh, Science, 40*(1), 141–150. https://doi.org/10.3329/jasbs.v40i1.31741.

Hartley-Whitaker, J., Ainsworth, G., Vooijs, R., Ten, B. W., Schat, H., & Meharg, A. A., (2001). Phytochelatins are involved in differential arsenate tolerance in *Holcus lanatus*. *Plant Physiology, 126*(1), 299–306. https://doi.org/10.1104/pp.126.1.299.

Hu, Z. Y., Zhu, Y. G., Li, M., Zhang, L. G., Cao, Z. H., & Smith, F. A., (2007). Sulfur (S)-induced enhancement of iron plaque formation in the rhizosphere reduces arsenic accumulation in rice (*Oryza sativa* L.) seedlings. *Environmental Pollution, 147*(2), 387–393. https://doi.org/10.1016/j.envpol.2006.06.014.

Jia, H., Ren, H., Gu, M., Zhao, J., Sun, S., Zhang, X., Chen, J., et al., (2011). The phosphate transporter gene ospht1;8 is involved in phosphate homeostasis in rice. *Plant Physiology, 156*(3), 1164–1175. https://doi.org/10.1104/pp.111.175240.

Kalita, J., & Tanti, B., (2020). Screening of some traditional rice cultivars of Assam, India, for their response to arsenic-induced abiotic stress. *Acta Agrobotanica, 73*(1), 1–16. https://doi.org/10.5586/aa.7315.

Kalita, J., Pradhan, A. K., Shandilya, Z. M., & Tanti, B., (2018). Arsenic stress responses and tolerance in rice: Physiological, cellular and molecular approaches. *Rice Science, 25*(5), 235–249. https://doi.org/10.1016/j.rsci.2018.06.007.

Kumar, S., & Trivedi, P. K., (2018). Glutathione S-transferases: Role in combating abiotic stresses including arsenic detoxification in plants. In: *Frontiers in Plant Science* (Vol. 9, pp. 1–9). https://doi.org/10.3389/fpls.2018.00751.

Li, G., Santoni, V., & Maurel, C., (2014). Plant aquaporins: Roles in plant physiology. *Biochimica et Biophysica Acta – General Subjects, 1840*(5), 1574–1582. https://doi.org/10.1016/j.bbagen.2013.11.004.

Li, R., Ago, Y., Liu, W. J., Mitani, N., Feldmann, J., McGrath, S. P., Ma, J. F., & Zhao, F. J., (2009). The rice aquaporin lsi1 mediates uptake of methylated arsenic species. *Plant Physiology, 150*(4), 2071–2080. https://doi.org/10.1104/pp.109.140350.

Ma, J. F., & Yamaji, N., (2015). A cooperative system of silicon transport in plants. *Trends in Plant Science, 20*(7), 435–442. https://doi.org/10.1016/j.tplants.2015.04.007.

Ma, J. F., Yamaji, N., Mitani, N., Xu, X., Su, Y., Mcgrath, S. P., & Zhao, F., (2008). Transporters of arsenite in rice and their role in arsenic accumulation in rice grain. *Journal of the American Medical Association, 105*(29), 9931–9935. https://doi.org/10.1001/jama.1896.02430750037005.

Mandal, B., & Suzuki, K., (2002). Arsenic round the world: A review. *Talanta, 58*(C), 201–235. https://doi.org/10.1016/S0065-2113(08)70013-0.

Meharg, A. A., & Jardine, L., (2003). Arsenite transport into paddy rice (*Oryza sativa*) roots. *New Phytologist, 157*(1), 39–44. https://doi.org/10.1046/j.1469-8137.2003.00655.x.

Mishra, S., Mattusch, J., & Wennrich, R., (2017). Accumulation and transformation of inorganic and organic arsenic in rice and role of thiol-complexation to restrict their translocation to shoot. *Scientific Reports, 7*(1), 1–13. https://doi.org/10.1038/srep40522.

Mitani, N., Yamaji, N., & Ma, J. F., (2008). Characterization of substrate specificity of a rice silicon transporter, Lsi1. *Pflugers Archiv. European Journal of Physiology, 456*(4), 679–686. https://doi.org/10.1007/s00424-007-0408-y.

Mohammed, A. K. S., Jayasinghe, S. S., Chandana, E. P. S., Jayasumana, C., & De Silva, P. M. C. S., (2015). Arsenic and human health effects: A review. *Environmental Toxicology and Pharmacology, 40*(3), 828–846. https://doi.org/10.1016/j.etap.2015.09.016.

Mondal, D., Rahman, M. M., Suman, S., Sharma, P., Siddique, A. B., Rahman, M. A., Bari, A. S. M. F., et al., (2021). Arsenic exposure from food exceeds that from drinking water in endemic area of Bihar, India. *Science of the Total Environment, 754*, 142082. https://doi.org/10.1016/j.scitotenv.2020.142082.

Moore, K. L., Chen, Y., Van De, M. A. M. L., Hughes, L., Liu, W., Geraki, T., Mosselmans, F., et al., (2014). Combined NanoSIMS and synchrotron X-ray fluorescence reveal distinct cellular and subcellular distribution patterns of trace elements in rice tissues. *New Phytologist, 201*(1), 104–115. https://doi.org/10.1111/nph.12497.

Mosa, K. A., Kumar, K., Chhikara, S., Mcdermott, J., Liu, Z., Musante, C., White, J. C., & Dhankher, O. P., (2012). Members of rice plasma membrane intrinsic proteins subfamily are involved in arsenite permeability and tolerance in plants. *Transgenic Research, 21*(6), 1265–1277. https://doi.org/10.1007/s11248-012-9600-8.

Murugaiyan, V., Zeibig, F., Anumalla, M., Siddiq, S. A., Frei, M., Murugaiyan, J., & Ali, J., (2021). Arsenic stress responses and accumulation in rice. In: Ali, J., & Wani, S. H., (eds.), *Rice Improvement: Physiological, Molecular Breeding and Genetic Perspectives* (pp. 281–313). Springer International Publishing. https://doi.org/10.1007/978-3-030–66530-2_9.

Nguyen, Q. T. T., Huang, T. L., & Huang, H. J., (2014). Identification of genes related to arsenic detoxification in rice roots using microarray analysis. *International Journal of Bioscience, Biochemistry and Bioinformatics, 4*(1), 22–27. https://doi.org/10.7763/ijbbb.2014.v4.304.

Norton, G. J., Islam, M. R., Duan, G., Lei, M., Zhu, Y., Deacon, C. M., Moran, A. C., et al., (2010). Arsenic shoot-grain relationships in field grown rice cultivars. *Environmental Science and Technology, 44*(4), 1471–1477. https://doi.org/10.1021/es902992d.

Panaullah, G. M., Alam, T., Hossain, M. B., Loeppert, R. H., Lauren, J. G., Meisner, C. A., Ahmed, Z. U., & Duxbury, J. M., (2009). Arsenic toxicity to rice (*Oryza sativa* L.) in Bangladesh. *Plant and Soil, 317*(1, 2), 31–39. https://doi.org/10.1007/s11104-008-9786-y.

Pandey, A. K., Gedda, M. R., & Verma, A. K., (2020). Effect of arsenic stress on expression pattern of a rice specific miR156j at various developmental stages and their allied co-expression target networks. *Frontiers in Plant Science, 11*, 1–11. https://doi.org/10.3389/fpls.2020.00752.

Pandey, S., Rai, R., & Rai, L. C., (2015). Biochemical and molecular basis of arsenic toxicity and tolerance in microbes and plants. In: *Handbook of Arsenic Toxicology*. Elsevier Inc. https://doi.org/10.1016/B978-0-12-418688-0.00027-7.

Pradhan, A. K., Jyoti, S. Y., Shandilya, Z. M., Rehman, M., Saikia, D., Poudel, J., & Tanti, B., (2021). Dissecting the molecular basis of drought-induced oxidative stress tolerance in rice. *Molecular Breeding for Rice Abiotic Stress Tolerance and Nutritional Quality*, 249–273.

Pradhan, A. K., Kalita, J., Lahkar, L., Gurung, L., Ghritlahre, S. K., & Tanti, B., (2020). Stress management in crops by utilizing landraces: Genetics and plant breeding perspective. In: *Sustainable Agriculture in the Era of Climate Change* (pp. 1–21). Springer, Cham.

Pradhan, A. K., Rehman, M., Saikia, D., Jyoti, S. Y., Poudel, J., & Tanti, B., (2020). Biochemical and molecular mechanism of abiotic stress tolerance in plants. In: *Plant Ecophysiology and Adaptation Under Climate Change: Mechanisms and Perspectives I* (pp. 825–853). Springer, Singapore.

Punshon, T., Jackson, B. P., Meharg, A. A., Warczack, T., Scheckel, K., & Guerinot, M. L., (2017). Understanding arsenic dynamics in agronomic systems to predict and prevent uptake by crop plants. *Science of the Total Environment, 581, 582*, 209–220. https://doi.org/10.1016/j.scitotenv.2016.12.111.

Raab, A., Williams, P. N., Meharg, A., & Feldmann, J., (2007). Uptake and translocation of inorganic and methylated arsenic species by plants. *Environmental Chemistry, 4*(3), 197–203. https://doi.org/10.1071/EN06079.

Rahman, I. M. M., & Hasan, M. T., (2007). Arsenic incorporation into garden vegetables irrigated with contaminated water. *Journal of Applied Sciences and Environmental Management, 11*(4), 105–112.

Rahman, M. A., Hasegawa, H., Rahman, M. M., Miah, M. A. M., & Tasmin, A., (2008). Straight head disease of rice (*Oryza sativa* L.) induced by arsenic toxicity. *Environmental and Experimental Botany, 62*(1), 54–59. https://doi.org/https://doi.org/10.1016/j.envexpbot.2007.07.016.

Seyfferth, A. L., Webb, S. M., Andrews, J. C., & Fendorf, S., (2011). Defining the distribution of arsenic species and plant nutrients in rice (*Oryza sativa* L.) from the root to the grain. *Geochimica et Cosmochimica Acta, 75*(21), 6655–6671. https://doi.org/10.1016/j.gca.2011.06.029.

Shi, S., Wang, T., Chen, Z., Tang, Z., Wu, Z., Salt, D. E., Chao, D. Y., & Zhao, F. J., (2016). OsHAC1;1 and OsHAC1;2 function as arsenate reductases and regulate arsenic accumulation. *Plant Physiology, 172*(3), 1708–1719. https://doi.org/10.1104/pp.16.01332.

Singh, V. P., Singh, S., Kumar, J., & Prasad, S. M., (2015). Investigating the roles of ascorbate-glutathione cycle and thiol metabolism in arsenate tolerance in ridged luffa seedlings. *Protoplasma, 252*(5), 1217–1229. https://doi.org/10.1007/s00709-014-0753-6.

Smedley, P. L., & Kinniburgh, D. G., (2002). A review of the source, behaviour, and distribution of arsenic in natural waters. *Applied Geochemistry, 17*(5), 517–568. https://doi.org/10.1016/S0883-2927(02)00018-5.

Song, W. Y., Park, J., Mendoza-Cózatl, D. G., Suter-Grotemeyer, M., Shima, D., Hörtensteiner, S., Geisler, M., et al., (2010). Arsenic tolerance in *Arabidopsis* is mediated by two ABCC-type phytochelatin transporters. *Proceedings of the National Academy of Sciences of the United States of America, 107*(49), 21187–21192. https://doi.org/10.1073/pnas.1013964107.

Song, W. Y., Yamaki, T., Yamaji, N., Ko, D., Jung, K. H., Fujii-Kashino, M., An, G., et al., (2014). A rice ABC transporter, OsABCC1, reduces arsenic accumulation in the grain. *Proceedings of the National Academy of Sciences of the United States of America, 111*(44), 15699–15704. https://doi.org/10.1073/pnas.1414968111.

Su, Y. H., McGrath, S. P., & Zhao, F. J., (2010). Rice is more efficient in arsenite uptake and translocation than wheat and barley. *Plant and Soil, 328*(1), 27–34. https://doi.org/10.1007/s11104-009-0074-2.

Suriyagoda, L. D., Dittert, K., & Lambers, H., (2018). Mechanism of arsenic uptake, translocation, and plant resistance to accumulate arsenic in rice grains. *Agriculture, Ecosystems, and Environment, 253*, 23–37. https://doi.org/10.1016/j.agee.2017.10.017.

Tang, Z., Chen, Y., Chen, F., Ji, Y., & Zhao, F. J., (2017). OsPTR7 (OsNPF8.1), a putative peptide transporter in rice, is involved in dimethylarsenate accumulation in rice grain. *Plant and Cell Physiology, 58*(5), 904–913. https://doi.org/10.1093/pcp/pcx029.

Verma, P. K., Verma, S., Pande, V., Mallick, S., Deo Tripathi, R., Dhankher, O. P., & Chakrabarty, D., (2016). Overexpression of rice glutaredoxin OsGrx_C7 and OsGrx_C2. 1 reduces intracellular arsenic accumulation and increases tolerance in *Arabidopsis thaliana*. *Frontiers in Plant Science*, *7*, 740. https://doi.org/10.3389/fpls.2016.00740.

Wang, F. Z., Chen, M. X., Yu, L. J., Xie, L. J., Yuan, L. B., Qi, H., Xiao, M., et al., (2017). OsARM1, an R2R3 MYB Transcription factor, is involved in regulation of the response to arsenic stress in rice. *Frontiers in Plant Science, 8*, 1–16. https://doi.org/10.3389/fpls.2017.01868.

Wenzel, W., (2013). Arsenic. In: Alloway, B. J., & Trevors, J. K., (eds.), *Heavy Metals in Soils*. Springer. https://doi.org/https://doi.org/10.1007/978-94-007-4470-7.

Williams, P. N., Price, A. H., Raab, A., Hossain, S. A., Feldmann, J., & Meharg, A. A., (2005). Variation in arsenic speciation and concentration in paddy rice related to dietary exposure. *Environmental Science and Technology, 39*(15), 5531–5540. https://doi.org/10.1021/es0502324.

Williams, P. N., Villada, A., Deacon, C., Raab, A., Figuerola, J., Green, A. J., Feldmann, J., & Meharg, A. A., (2007). Greatly enhanced arsenic shoot assimilation in rice leads to elevated grain levels compared to wheat and barley RID A-4269-2009 RID B-7917-2008 RID A-5189-2008 RID B-8079-2011. *Environmental Science & Technology, 41*(19), 6854–6859.

Wu, Z., Ren, H., Mcgrath, S. P., Wu, P., & Zhao, F. J., (2011). Investigating the contribution of the phosphate transport pathway to arsenic accumulation in rice. *Plant Physiology, 157*(1), 498–508. https://doi.org/10.1104/pp.111.178921.

Xie, M. Y., Tian, Z. H., Yang, X. L., Liu, B. H., Yang, J., & Lin, H. H., (2019). The role of OsNLA1 in regulating arsenate uptake and tolerance in rice. *Journal of Plant Physiology, 236*, 15–22. https://doi.org/10.1016/j.jplph.2019.02.013.

Xu, J., Shi, S., Wang, L., Tang, Z., Lv, T., Zhu, X., Ding, X., et al., (2017). OsHAC4 is critical for arsenate tolerance and regulates arsenic accumulation in rice. *New Phytologist, 215*(3), 1090–1101. https://doi.org/10.1111/nph.14572.

Xu, X. Y., McGrath, S. P., & Zhao, F. J., (2007). Rapid reduction of arsenate in the medium mediated by plant roots. *New Phytologist, 176*(3), 590–599. https://doi.org/10.1111/j.1469-8137.2007.02195.x.

Yang, H. C., Fu, H. L., Lin, Y. F., & Rosen, B. P., (2012). Pathways of arsenic uptake and efflux. In: *Current Topics in Membranes* (Vol. 69). Elsevier. https://doi.org/10.1016/B978-0-12-394390-3.00012-4.

Ye, Y., Li, P., Xu, T., Zeng, L., Cheng, D., Yang, M., Luo, J., & Lian, X., (2017). Ospt4 contributes to arsenate uptake and transport in rice. *Frontiers in Plant Science, 8*, 1–12. https://doi.org/10.3389/fpls.2017.02197.

Zhang, J., Martinoia, E., & Lee, Y., (2018). Vacuolar transporters for cadmium and arsenic in plants and their applications in phytoremediation and crop development. *Plant and Cell Physiology, 59*(7), 1317–1325. https://doi.org/10.1093/pcp/pcy006.

Zhao, F. J., Ago, Y., Mitani, N., Li, R. Y., Su, Y. H., Yamaji, N., McGrath, S. P., & Ma, J. F., (2010). The role of the rice aquaporin Lsi1 in arsenite efflux from roots. *New Phytologist, 186*(2), 392–399. https://doi.org/10.1111/j.1469-8137.2010.03192.x.

Zhao, F. J., Harris, E., Yan, J., Ma, J., Wu, L., Liu, W., McGrath, S. P., Zhou, J., & Zhu, Y. G., (2013). Arsenic methylation in soils and its relationship with microbial arsM abundance and diversity, and as speciation in rice. *Environmental Science and Technology, 47*(13), 7147–7154. https://doi.org/10.1021/es304977m.

Zhao, F. J., Ma, J., Meharg, A., & McGrath, S., (2009). Arsenic uptake and metabolism in plants. *New Phytologist, 181*, 777–794.

Zhao, F. J., Moore, K. L., Lombi, E., & Zhu, Y. G., (2014). Imaging element distribution and speciation in plant cells. *Trends in Plant Science, 19*(3), 183–192. https://doi.org/10.1016/j.tplants.2013.12.001.

CHAPTER 7

Agronomic Measures to Lessen Arsenic in Rice

ANSHIKA KAUSHIK,[1] AMARNATH MISHRA,[2] VARAD NAGAR,[3] PRITAM P. PANDIT,[3] RUSHIKESH L. CHOPADE,[3] ROHIT KUMAR VERMA,[4] APOORVA SINGH,[3] and SHIKHA VERMA[5]

[1]*Department of Anthropology, University of Delhi, India*

[2]*Amity Institute of Forensic Sciences, Amity University, India*

[3]*Department of Forensic Science, Vivekananda Global University, India*

[4]*Dr. APJ Abdul Kalam Institute of Forensic Science and Criminology, Bundelkhand University, India*

[5]*French Associates Institute for Agriculture and Biotechnology of Drylands, The Jacob Blaustein Institutes for Desert Research, Ben-Gurion University of the Negev, Sede Boqer Campus, Israel*

ABSTRACT

Soils that are used for rice cultivation comprise high concentrations of arsenic. Extremely use of agrochemicals or industrial residues containing arsenic in many areas and irrigation with groundwater containing As are major factors that contribute to arsenic concentration in soil. For the reduction of arsenic into rice through soils or its uptake, rice cultivation under aerobic conditions, intermittent flooding conditions, rather than in flooded soils are done. The use of phosphorous, iron, silicon, and different fertilizers has to be practiced carefully or should be reduced manually because their effects have shown negative results. Also, the use of arsenic

Arsenic in Rice: A Global Menace. Garima Awasthi, Sudhakar Shrivastava, and Mahipal Singh Sankhla, (Eds.)

hyperaccumulating plant species individually or with another plant type (e.g., *Azolla*, *Chlorella*, or *Nannochloropsis* species) helps to reduce arsenic in rice. To minimize arsenic uptake in rice from soils, bioavailability is considered. But factors like arsenic speciation under different conditions such as examining them for a long-time, in different field settings are required prior to the recommendation of methods in farming.

7.1 INTRODUCTION

In the modern era, rice-producing regions around the world are increasingly being affected by arsenic contamination, which can have serious health effects. Arsenic accumulates in rice grains, and this can lead to major problems for human health. South and Southeast Asia are the most affected. The quantity of arsenic in the field soil has a negative impact on normal growth, plant metabolism, and plant development, which ultimately leads to a decrease in the yield of many crops, including rice plants (Gupta et al., 2017).

Research shows that chronic arsenic poisoning causes serious health consequences, including cancer, Blackfoot disease, hardening of the skin, gangrene, restrictive lung disease, diabetes mellitus, hypertension, and other heart conditions. The exposure of humans to arsenic occurs by different means, such as food, water, soil, dust inhalation, and majorly by consumption of rice, which is considered the main source of arsenic, especially in the tropics (Gupta et al., 2017; Martinez et al., 2011).

Human activities such as using contaminated water for irrigation, digging up the earth for groundwater, misapplication of arsenic-containing agrochemicals such as aromatic arsenic, and mining of additives to increase arsenic in agricultural soils are unanimous (Sahoo et al., 2013). It is considered an environmental toxin because it is useless for living organisms, although arsenic, with the symbol As, is one of the most copious elements in the earth's crust and its occurrence is natural in the environment (Gupta et al., 2017; Fahad et al., 2015).

Arsenite, arsenate, dimethyl arsenic acid, monomethyl arsenic, and methylarsonic acid are the inorganic forms of arsenic. Other organic forms, such as trimethylarsine oxide (TMAO), arsenobutin (AB), arcinocholine (AC), dimethyl arginyl riboside, glyceryl phosphorous larcinocholine are also present in the environment but in low concentrations (Fahad et al.,

2015; Banerjee et al., 2013). The organic forms of arsenic are not a major component in soil, while the inorganic forms such as As(III) and As(V), are the predominant forms extending from rice roots from contaminated soils (Bhattacharya et al., 2014).

7.2 HEALTH EFFECTS OF ARSENIC FROM RICE DIET ON HUMANS

Rice is the main source of meals in many countries, mostly in Asian and Southeast Asian countries. Consumption of vegetable-rich foods lowers the risk of arsenic contamination and its related health issues due to different rates of arsenic metabolism. Moreover, our intestinal microbial play a major role in arsenic speciation and its absorption into the blood. It undergoes a series of bio-transformation inside our gastrointestinal tract, including oxidation, reduction, and methylation. The pH of our stomach is acidic which adds to the increasing arsenic bio-accessibility in comparison to the intestine. Much research shows that the method of food processing also decides the exposure to arsenic through the meals like the temperature, duration, and the way and medium of cooking. Well-cooked rice with plentiful water reduces arsenic concentration in rice (Farooq et al., 2016; Li et al., 2009).

The bioavailability of arsenic, root uptake, accumulation, transport, and grain delivery of arsenic are vital aspects of research to ease arsenic in rice (Table 7.1).

7.3 AGRONOMIC APPROACHES FOR REDUCING ARSENIC ACCUMULATION IN RICE

Among the naturally arising metalloid in the soil, arsenic is one of them. It is present in some areas in large quantity which is due to human activity such as the use of pesticides that contains arsenic. The factors that include the oxidative pressure in rice, its translocation, and biotechnological approaches to decrease uptake, are important to know in agronomic strategies and, thus the accumulation of arsenic content in rice has been studied (Azam et al., 2014). Many agronomic strategies or methods which decrease the lethargy of arsenic accumulation in rice, include:

TABLE 7.1 Listing Various Health Hazards Due to Consumption of Arsenic-Contaminated Rice

Effect on Organ System	Exposure	Symptoms	Fatal Dose	References
Respiratory effects	Arsenic dust and fume inhalation in the mining of ores in the industrial area during smelting.	Irritation of mucous membranes can lead to laryngitis and bronchitis. Block nose, sore throat, hoarseness of voice, and chronic cough. Arsenic trioxide in lethal condition causes bronchial mucosa, sub-mucosal hemorrhages, and pulmonary edema. Asthmatic bronchitis and asthma are common in groundwater arsenic toxicity.	0.01–1 mg/m^3	Wu et al. (2011); Paszkowski et al. (2002); Kamiya et al. (2013)
Cardiovascular effects	Arsenic ingestion through food or water.	Damage the blood vessels of the heart. Myocardial depolarization which ultimately leads to the failure of the heart. Vascular system damage, an example of which is Blackfoot disease is seen in case of low-level exposure.	0.5 to 2.45 mg/m^3	Li et al. (2009); Wang et al. (2016); Ma et al. (2008)
Gastrointestinal effect	Consumption of tube wells water having arsenic content.	Nausea, sickness, vomiting, and diarrhea. Clinical signs such as gastrointestinal irritation, burning of lips, pain during swallowing, feeling thirst, and nausea from acute arsenic poisoning are seen.	64 mg/m^3	Huang et al. (2012); Mitra et al. (2014)
Hematological effects	Short-term and long-term arsenic exposures. Mostly oral exposures.	Anemia, leukopenia, and direct hemolytic or cytotoxic effects can be seen in the blood cells. High concentrations of arsine (10 ppm) cause death due to hemolysis of red blood cells. And renal damage is secondary.	0.2 mg/m^3	Zhao et al. (2010); Raab et al. (2007)
Hepatic effect	Oral exposure	Bleeding in the esophagus, jaundice, and liver become enlarged. Formation of a Hepatic lesion after persistent ingestion of arsenic-containing medicines for a long duration.	0.02 to 0.1 mg/m^3	Gonzalez et al. (2014)

TABLE 7.1 *(Continued)*

Effect on Organ System	Exposure	Symptoms	Fatal Dose	References
Renal effects	Oral exposure	Arsenic poisoning produces shock and leads to dehydration, a high risk of renal failure. In some cases, dialysis has been effective in overcoming the problem. Hemolysis induced by arsenic cause tubular necrosis with partial or complete renal failure.	10–15 mg/m^3	Norton et al. (2010); Carey et al. (2010); Zhao et al. (2008)
Dermal effects	Drinking water that contains arsenic in trace amounts.	Hyperkeratosis is the formation of warts or corns on the palms and soles of hands, and hyperpigmentation intermixed with small areas of hypopigmentation on the neck, face, and back.	0.01 to 0.1 mg/m^3	Shi et al. (2016); Jia et al. (2013)
Neurological effects	Ingestion of inorganic arsenic.	Damaging of the cardiovascular system and the nervous system. A neurological injury like Coma, headache, confusion, hallucination, seizures, and lethargy, causes numbness, loss of reflexes, and muscle weakness.	1 mg/m^3	Song et al. (2014)
Reproductive effects	Human oral exposure, occupational, and medical exposure.	Damage of DNA and genetic alterations in wide ranges, which are from simple gene mutations to large visible changes in chromosome structure. These changes in excess may cause genetic damage transferable to succeeding generations leading to carcinogenic effects, skin, and lung cancer, etc.	22 mg/m^3	Xu et al. (2015); Hernandez et al. (2015); Meharg et al. (2012)

- Providing suitable condition that promotes the development and precipitation of insoluble arsenic in soil;
- Providing air in the soil by water running and thus checks the reduction of arsenic in rice;
- Reduction in arsenic uptake and its translocation in plants by providing more and more minerals in the soil that will compete with arsenic uptake (Zheng et al., 2013; Pinson et al., 2015).

Many effective remedies that contribute to decreasing the risk related to arsenic in plants are:

- Impregnation of soil with minerals;
- Bioremediation strategy;
- Providing more water management and changing irrigation practices.

7.3.1 IMPREGNATION OF SOIL WITH MINERALS

Some specific mineral nutrients like Fe, S, P, and Si lessen the accumulation of arsenic in plants. Thus, incorporating these minerals will decrease arsenic accumulation in edible parts by targeting its uptake and its translocation in food crops (Bastias et al., 2016).

7.3.2 ROLE OF FE

The presence of iron within the plant's vascular system leads to the release of iron oxide, which coats the rice plants' roots and reduces arsenic absorption. Ferrum is an essential plant mineral nutrient. Arsenic reduction is essential for reducing the accumulation of arsenic in rice. Co-precipitation of iron occurs in rice plants and arsenic too. This means that iron is absorbed onto the rice surface, reducing the presence of soluble arsenic in rice plants.

During rice cultivation, anaerobic conditions promote the formation of iron plaques throughout the roots, and the iron plaques are mainly composed of hydrated iron ore, siderite, and goethite (Dalgedo et al., 2016). Iron spot plays an important role in reducing the uptake of arsenic in rice because it has a high affinity for arsenic and it will seize arsenic, which will ultimately lead to reduced arsenic transport from roots to shoots. The concentration of Ferrum oxides rises in the rhizosphere of rice plants, which will subsequently decrease arsenic uptake in the plants (Figure 7.1).

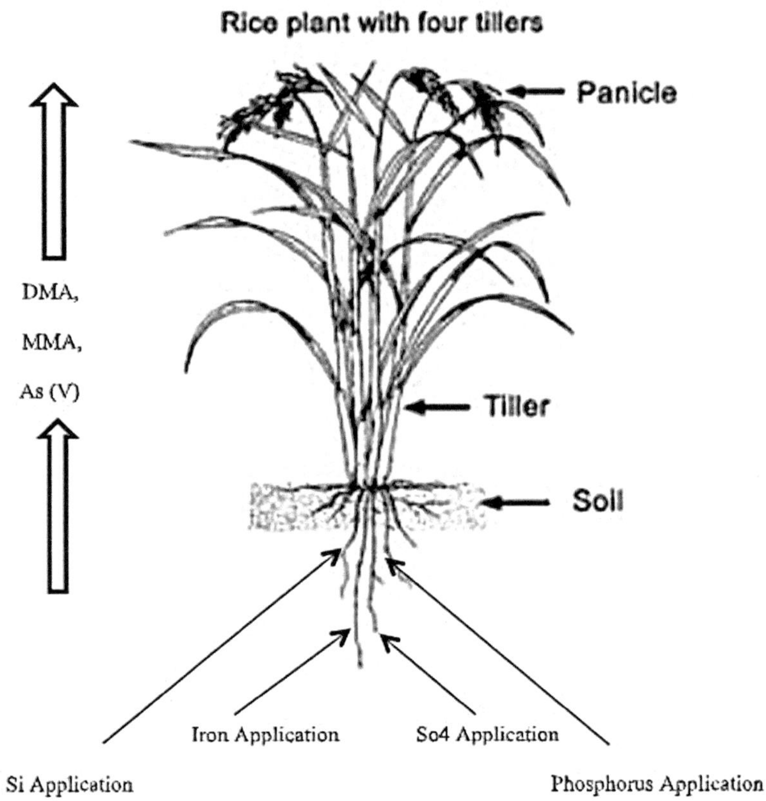

FIGURE 7.1 Diagrammatic representation of uptake of arsenic in rice plant and application of Si, Fe, P, and SO_4.

Figure 7.1 represents the application of Fe, P, S, and Si through its roots for its remediation and to minimize the uptake of arsenic concentration as much as possible in the rice plants. On the other side, the uptake of both forms of arsenic be it organic or inorganic, and its accumulation in the rice grains are shown.

The application of Fe-oxide has shown better results in reducing arsenic concentration in rice grains in rice fields. Arsenic uptake is majorly governed by the bioavailability of arsenic rather than the total arsenic concentrations present in the soil. Arsenic binds with Fe-oxides in the soil thus reducing arsenic movement in soil solution. Oxides of Fe act as a fall for arsenic, thus, the higher the Fe oxide in soil lower the uptake less the accumulation of arsenic in rice plants (Islam et al., 2004; Marim et al., 1992).

7.3.3 ROLE OF PHOSPHORUS

Another important nutrient for arsenic solubility in soil is phosphorus. Soil uses phosphate instead of arsenic to compete with arsenate. The incorporation of phosphate into the soil has resulted in reduced arsenic levels in Fe plaques, leading to an increase in arsenic solubility. At critical concentrations, phosphate exerts an inhibitory effect, competing with arsenate for the same carrier during uptake. The phosphate concentration in the solution increases, which reduces the uptake of arsenate by the plant (Williams et al., 2007; Lauren et al., 2005).

One important factor in soil control is the effect of phosphorus on the movement of arsenic. The factors that affect the uptake and absorption of rice are as follows:

- For adsorption sites on soil particles, the antagonistic behavior between arsenic and phosphorous.
- From root to shoot, the role of Pi in transporting arsenic.
- During absorption from rice roots, the role of Pi in transporting arsenic from root to stem, inorganic phosphate (Pi) shows an anti-arsenic effect (Takahashi et al., 2004; Kraemer et al., 2004).

7.3.4 ROLE OF SILICA

Plants can use only mono-silicic acid among the diverse soluble Silicon forms that exist in the soil. Supplementation of silica improves crop yield by improving the number of spikelets per panicle. pH controls the solubility of Silica. Also, silica helps in enhancing the primary metabolism and thus stimulates the remobilization of amino acids in rice (Bhattacharya et al., 2008).

The higher concentration of silica in the soil lowers arsenite uptake by rice. Both arsenite and silicon compete during the uptake. Silica supply decreases the expression of the silica transporters Lsi1 and Lsi2, for the other mediators of arsenite uptake arsenic concentration in rice plantlets. Among silica concentration and uptake of inorganic arsenic (Asi) species extensive destructive connection was studied (Adriano et al., 2001).

Some parts were unenviably affected by arsenic exposure in the absence of silica supplementation like the growth of shoots. Low uptake of arsenic took place in the presence of silica and it triggered the enhancement in the antioxidant defense response in plants. Silica proved to be more beneficial in decreasing arsenic concentration in leaves, flags, husks, straws, and grains of rice. The arsenic accumulation in rice straw and grains showed reduced arsenic concentration in leaves due to the solicitation of silica in soil (Quazi et al., 2011; Chatterjee et al., 2013; Signes-Pastor et al., 2007).

7.3.5 ROLE OF SULFUR

The role of sulfur in reducing the accumulation of arsenic and translocation in plants is mentioned below. Sulfur is an essential element required for the growth of plants.

Mostly by three methods sulfur considerably reduces arsenic accumulation or concentration in rice:

1. Sulfur enhances the reduction of arsenate from Fe-plaques;
2. Sulfur persuades the Fe plaques establishment on the surface of the roots, and rhizosphere of the rice plant, and arsenic concentration in the soil decreases;
3. Transport site SO_4 hinders arsenate transport into cells at the cell membrane (Campbell et al., 1985; Ahmed et al., 2011).

Arsenic exposure to plants promotes the synthesis of low molecular weight sulfur-rich ligands such as glutathione and phytokeratin (Chatterjee et al., 2013). Sulfur metabolism plays an important role in the detoxification process. It is important for plants to survive in arsenic-contaminated soil. Incorporating sulfate (SO4) into rice soil has another benefit. Sulfates have a strong affinity for arsenic under reducing conditions and precipitate as insoluble arsenic sulfide (Williams et al., 2011; Pikaray et al., 2005; Rahaman et al., 2011). In laboratory experiments, arsenic concentration in rice sprouts decreased with increasing S intake. This was due to S-induced Fe plaque formation (Hu et al., 2007; Fan et al., 2013). In contrast, Tripathi, etc. (2014) It was found that Fe plaque formation induced by excessive S supply is a possible explanation for the increase in arsenic concentration in rice roots. Therefore, it is hypothesized that the decrease in arsenic concentration in rice may be due to the high arsenic complex formation in rice roots in soils with high S availability (Zhang et al., 2011). Recently, Dixit et al. (2015) also reported an increase in arsenic accumulation in rice roots after high sulfur treatment, resulting in a decrease in arsenic root-to-shoot translocation. This was due to the induced synthesis of thiols and the resulting higher arsenic complexion in the roots of rice. However, this hypothesis has not been tested in this area and/or in the future. Therefore, given the availability of supplements containing S, further field trials are needed to reduce the concentration of arsenic in rice grains.

7.3.6 SILICON

The capacity of silicon to enrich toxicity (metallic) has been assessed more and more at the current time (Meharg & Meharg, 2015; Sanglard et al., 2016). As(III) and acid each possess an excessive pKa 1 (9.2–9.5) and have tetrahedral structures with comparable size, As(III) shares the silicon delivery structures for uptake in root cells of rice (Guopet et al., 2007; Ma et al., 2008; Zhao et al., 2010). Subsequently, Silicone, and As(III) may also compete for the above-mentioned part. Apart from the negative effects, Si influences a variety of minerals of arsenic in the stable stage of field soil, i.e., the application of silicon lowers the arsenic proportion which is found to be associated with iron and aluminum hydrous oxides (Wu et al., 2016). A number of researches proved that extra Silicon, both in aquiculture or in soil, lowers the transportation of inorganic kinds of arsenic from shoot to the rice grain (Bogdan & Schenk, 2009; Li et al., 2009; Seyffreth & Fendrof, 2012; Fleck et al., 2013; Liu et al., 2014; Yu et al., 2016; Wu et al., 2016). The inorganic kinds of arsenic may be decreased by means of including silicate gel that's soluble in paddy soils, however, heightened the transportation of DMA (Li et al., 2009; Liu et al., 2014). It has been advised that an improvement in the uptake process of DMA is resulting from dissolved acid due to silicate gel surpassing

DMA for soil, growing its absorption in the soil (Liu et al., 2014). The universal impact of Silicone fertilizers or their software in soils to lessen arsenic uptake by means of rice, i.e., arsenic awareness may also get heightened if reduced costs of Silicone are introduced to the field soil which is deficient in silicone. Moreover, toxicity and uptake of arsenic are probably assuaged if adjustments with excessive Silicone concentration are supplemented. Hence, the worldwide reaction of Si in the direction of decreasing arsenic toxicity is determined by means of the portions of Si and arsenic gift withinside the soil, an amount of Si supplemented by way of fertilizers, and soil biogeochemistry. Among the various mitigation approaches, the application of fertilizers constituting Silicone is a really favorable method and could have primacy in destiny studies in phrases of As concentration in rice. The mechanism involved in arsenic transportation and plant resistance in the direction of growing arsenic in grains of rice has currently been reviewed by means of numerous researchers (Punshon et al., 2017; Suriyagoda et al., 2018). The variety of rice with advanced characteristics such as radial oxygen loss, root porosity, or development of iron plaques which bind extra on Fe plaques, plummeting arsenic uptake (Pan et al., 2014). In arsenic-infected rice soil, such cultivars show to be a resource in decreasing the awareness of arsenic as radial oxygen loss is the main process attributable to oxidization of rhizosphere, persuading improvement of iron plaques further adapting pretty soluble As(III) to less soluble As(III). Arsenic toxicity is in addition reduced by means of the manufacturing of ROS and antioxidant enzymes (Finnegan & Chen, 2012; Verma et al., 2020). The discount of arsenic concentrations in rice plant life follows the order of roots > leaves > grains. These arsenic escaping and countering techniques are affected by environment x Control interactions x cultivar. Furthermore, more differences in rice grain arsenic awareness amongst rice varieties and geographical aspects are said in exclusive studies (Arunakumara et al., 2013; Suriyagoda et al., 2018). Therefore, the selection of crops is a completely critical assignment meant to lessen arsenic awareness in rice grains.

7.3.7 NITROGEN FERTILIZATION

The development of Iron plaques on roots and reductive dissolution of iron in its oxide or hydroxides form in field soil cultivating rice are motivated with the aid of using nitrogen, therefore, affecting the uptake of arsenic in rice. Though Iron plaque development on the basic layer became reduced, nitrate-established stimulation oxidation of Fe(II) and further oxidation/ inhibition of Fe (III) within the bulk soil can also additionally sequester cellular arsenic in soil, ultimately ensuing in dropped arsenic awareness in rice crop (Chen et al., 2008). On the contrary, including ammonium stepped forward Fe(III) discount. Contempt those responses, below area situations over 4 cropping seasons in Hunan, China, involving nitrogen that had various consequences on arsenic accretion and speciation in rice grains (Ma et al., 2014). Furthermore, the application of nitric oxide to aquicultural ought to inverse (simplest partially) arsenic-brought about the upsurge in moves of different enzymes such as antioxidant, superoxide dismutase, catalases, peroxidases, etc., and expression stage of transporter such as OsLsi1 which is responsible for uptake of As(III) (Singh et al., 2009).

7.3.8 ORGANIC MATTER

Several studies reported a noteworthy indirect association between organic matter present in soil and the concentration of arsenic rice grains (Das et al., 2008; Fu et al., 2011; Rahaman et al., 2011; Kar et al., 2013). Organic acids contribute to lowering the transportation of arsenic in two ways either it develops complexes that are insoluble in nature or sometimes act as a binding agent (Wang & Mulligan, 2006). The decomposing organic matter in soil develops dissolved organic carbon which can bring arsenic existing in the topmost layers to deeper layers through percolation, deeper layers through percolation, thereby reducing arsenic concentration within the zones of root (Farooq et al., 2010).

High soil acidity reduces arsenic uptake into plants (Bogdan & Schenk, 2009; Chen et al., 2012; Desplanques et al., 2006; Guo et al., 2005; Ma & Takahashi, 2002; Li et al., 2009; Seyffreth & Fendorf, 2012; Seyfferth et al., 2016). Acids and As(III) have chemically similar structures and therefore compete with each other for root uptake. The reduced uptake of As(III) by rice in the occurrence of higher concentrations of acid is because of the reduced expression of Lsi1 and Lsi2 (Ma et al., 2006, 2007). Lsi1, a transporter of NIP, facilitates the transportation of Silicone via the surface medium of the roots of rice plants (Ma et al., 2006, 2008; Verma et al., 2016a).

Lsi1 facilitates the transportation of Arsenite from the soil to the roots of the plant (Ma et al., 2008). Consistent negative relationships between soil Si and rice root arsenic bioavailability were found in all studies, but most of the results were produced using controlled field settings (Bogdan & Schenk, 2009; Fleck et al., 2013; Li et al., 2009; Liu et al., 2014; Seyfferth & Fendorf, 2012; Wu et al., 2016; Yu et al., 2016). It is significant to assess the impact of Silicone on arsenic and the dynamics of rice fields on a long-term basis.

7.3.9 ROLE OF SOIL MICROORGANISMS

The quantity of minerals in the field soil is under the control of various soil microorganisms through various mechanisms like immobilization and mineralization that influence the transport of As in the rice plants. Decontamination of arsenic species by soil microorganisms through sorption is done. The microorganisms' extracellular surface amino sugars and proteins which has hydrogen bonding ability (Fu et al., 2011). *Bacillus* species, *Rhodococcus* species, *Halobacterium* species, and many other soil bacteria (Meharg et al., 2003; Bhattacharya et al., 2010) are found to adsorb various species of arsenic. The formation of liquid Iron hydroxides on the surface of the cell through the development of internal-sphere complexes is the alternative possible method for detoxification of As in soil microorganisms (Meharg et al., 2003; Bhattacharya et al., 2010; Zhang et al., 2008).

A number of studies have been performed to investigate the effect of microbes present in the rhizosphere on arsenic biotransformation and bioavailability in rice plants. In the studies which are conducted in laboratories, it was proved that oxidation, respiration of both arsenite and arsenate along with methylation of arsenite can be influenced by microorganisms (Verma et al., 2016b; Xiao et al., 2016; Malasarn et al., 2004). A recent study by Yang et al. (2015) demonstrated that Brevibacillus laterosporus may be accountable for the accelerated oxidation of arsenite, further inoculation of soil with B. laterosporous could be considered as an effectual way to lessen the noxiousness of As(III) in rice plant. Moreover, the efficacy of B. laterosporous in the oxidation process of As(III) in flooded soils would be contingent significantly on its rhizosphere via ROS. However, the capability of such bacteria in terms of oxidizing As(III) is expected to be more in rice culture which is under aerobic conditions. The genera Anaeromyxobacter, Desulfuromonas, Desulfobulbus, Desulfocapsa, Geobacter, Lacibacter, and Ohtaekwangia were present in the comparatively highly profused area in flooded soils, wherein in non-flooded one Acinetobacter, Ignavibacterium, Lysobacter, and Thiobacillus were the dominated ones. The methylation of arsenic species is catalyzed by arsM (arsenite methyltransferases), further attributable to different soil microorganisms (Qin et al., 2006; Ye et al., 2012). The microbes present in the field soil having arsM (gene) are diverse in phylogenetic aspect (Jia et al., 2013). The process of methylation of arsenic in soil is enriched by various factors that endorse anaerobic conditions (Mestrot et al., 2009; Norton et al., 2013; Jia et al., 2013). In a study that compare different soils, it was found that the quantity of arsenic species which are methylated is correlated with the various soil conditions like pH and therefore the concentration of DOC (Zhao et al., 2013; Ma et al., 2014). Recent research commends that the plants are unable to methylate arsenic on their own (Lomax et al., 2012), hence all methylated species of As within plants are possibly resulted as a result of environmental conditions, including the soil microflora (Arao et al., 2011a; Lomax et al., 2012). Arsenic can be transformed into its form which is having a smaller amount of toxicity, is more soluble, and mobile, along with bioavailable forms via metal-tolerant bacteria promoting plant growth further, contributing to the improvement of plant growth by producing constituents such as siderophores, hormones, enzymes, etc., that are involved in promoting the growth of plants (Wani et al., 2007; Singh et al., 2016a, b). Bacteria play manifold roles in arbitrating stress tolerance and modulating the translocation of edible parts of rice plants. Though, the influence of microflora in the reduction of arsenic toxicity in rice plants by means of the mentioned measures may be a temporally and spatially active method suffering from both management of crop and soil characteristics. Many studies proved the involvement of AM (arbuscular mycorrhizal) fungi in the uptake of arsenic via various plants (A1-Agely et al., 2005; Trotta et al., 2006; He & Lilleskov, 2014) which may also impact the dynamics of arsenic in rice soil (Li et al., 2011; Chen et al., 2012; Batista et al., 2016).

7.3.10 CONSTRAINT OF ARSENIC AT UNDERGROUND LEVEL

Increasing the level of arsenite-thiol complexation will reduce arsenic in rice grains, thus phytostabilizng the metalloid in the inedible parts or the underground biomass. Increased formation of GSH and PC will lead to the restraint of arsenic in rice plants (Norton et al., 2009).

7.3.11 INCREASE AS(III) RATE

Rice plants that are transgenic showed arsenic concentration to be low in the shoot and root part as compared to the wild plants which are having the same factor for arsenic translocation. These rice plants which expressed a type of ACR3 gene encoding As(III) efflux protein get enhanced and reduces arsenic accumulation in the rice grains (Liu et al., 2004; Duan et al., 2011).

7.3.12 TEXTURE

Arsenic is attracted and may bind to hydrous surfaces such as clays containing a high percentage of Fe, Mn, and aluminum oxides and get bind to them (Sadiq, 1997). Thus, soil with greater magnitudes of clay/ or hydrous area may have more arsenic in turn plummeting its bioavailability. Arsenic dynamics is affected by various factors like biological, chemical, and physical characteristics properties of soil along with arsenic concentration and thus the bioavailability and quantity within the solution of soil. Furthermore, despite the complete consideration that the incidence of increased phosphorous and iron within soil decreases the bioavailability of arsenic in soil and hence the organic matter which is present in high quantity might upsurge arsenic availability in soils contaminated with arsenic, contrary responses also can be anticipated, as seldomly reported, this might flow from the interactions which are dynamic and complex in nature influential factors in soil (Liu et al., 2004; Geng et al., 2005; Furrow et al., 2015). The comparative variations in these persuasive factors need to be considered together so as to extract eloquent inferences from such intricate conditions aimed toward assuaging arsenic toxicity in rice.

7.3.13 BIOCHAR APPLICATION

The pyrolysis of animal and plant remains under hypoxia is called pyrolysis (Wang et al., 2019). It often consists of a scattering of biomass waste (straw, straw, sawdust, mud, dregs, etc.) (Zheng et al., 2010). It has many advantages such as it is cheap and efficient, durable, has oxygen-rich functional groups, developed structure of pore with specifically large surface area, also has excellent potential for applications in contaminated water and soil remediation (Dahlawi et al., 2018b; Che et al., 2019; Dong et al., 2019; Yang et al., 2019). The application of biochar to reduce arsenic in rice soils appears to be a novel approach, although this method has also been verified for other benefits, primarily for upland crops (Lehmann, 2007; Woolf et al., 2010; Agrafioti et al., 2014). From the environmental point of view, the high porosity and the surface of the biochar make it an excellent means of remediation and remediation of soils contaminated by organic and inorganic pollutants (Arao et al., 2011b; Tang et al., 2013; Cabrera et al., 2014; Cope et al., 2014).

Pyrolysis of animal and plant remains in hypoxia is called pyrolysis (Wang et al., 2019). It often consists of a spread of biomass waste materials (straw, chaff, sawdust, sludge, lees, etc.) (Zheng et al., 2010). It has numerous advantages such as economic efficiency, and durability, contains rich oxygen-containing functional groups, developed pore structure with large specific surface and deep surface charge, also has excellent potential for contaminated water and soil remediation applications (Dahlawi et al., 2018b; Chen et al., 2019; Dong et al., 2019; Yang et al., 2019). The application of biochar to reduce arsenic in rice soils is emerging as a new approach, although this method has also been verified for other benefits, mainly for mountain crops (Lehmann, 2007; Woolf et al., 2007; Agrafioti et al., 2014). From an ecological point of view, the high porosity and large surface area of biochar make it an excellent tool for the restoration and remediation of soils contaminated by organic and inorganic pollutants (Arao et al., 2011b; Tang et al., 2013; Cabrera et al., 2014; Cope et al., 2014).

Much research has suggested the practice of biochar as a sorbent for metals from contaminated water and soil (Agrafioti et al., 2014) and has focused on the immobilization of metal cations such as lead

(Pb), copper (Cu), nickel (Ni) and cadmium (Cd), with few studies on arsenic removal by biochar and conflicting results reported (Beesley et al., 2011; Beesley & Marmiroli, 2011; Marmiroli, 2011; Agrafioti et al., 2014; Ma et al., 2014). Despite the discrepancies, the application of biologically modified biochar, such as CaO-saturated or Fe(III)-treated rice hulls, or Mn oxide-modified biochar, showed higher arsenic soaking efficiencies than the use of biochar derived solely from agricultural by-products (Samsuri et al., 2013; Agrafioti et al., 2014; Yu et al., 2017).

However, the application of biochar to release arsenic-contaminated irrigation water previously applied to rice soils would be a potentially effective method to eliminate the contribution of arsenic to rice fields. In addition, changing biochar can inflate the application of biochar as well as its scale and performance, and loading nanometal particles or metal oxide particles on the surface of biochar is a very effective modification method. Currently, iron-enriched biochar is widely used for arsenic removal in solution because arsenic has a strong affinity for Fe oxides (Yang et al., 2015; Wu et al., 2018). Agrafioti et al. (2014) found that As5+ and Cr6+ exclusion rates were enhanced by the application of biochar supplemented with calcium (Ca2+) and iron (Fe0, Fe3+) reagents. Lin et al. (2017) found that ferromanganese binary oxide not only facilitates the oxidation of As(III) to As(V) but also reduces the bioavailability of arsenic and shortens its accumulation in rice. Qiao et al. (2018) applied zero-valent iron (ZVI) biochar improvers to soil contaminated with arsenic and Cd, and thus the result showed that ZVI biochar was able to simultaneously reduce the bioavailability and mobility of arsenic and Cd in soil. Wu et al. (2018) compared the remediation effect of Fe-oxyhydroxy sulfate (Biochar-FeOS), FeCl3 (Biochar-FeCl3), and zero-valent iron (Biochar-Fe) and reported that biochar-FeOS has the main effective immobilization for the remediation of land contaminated by arsenic. The stabilization mechanism of soil arsenic by iron-modified biochar mainly involves changes in soil pH, natural processes, redox reaction, and adsorption (Agrafioti et al., 2014; Yang et al., 2015; Qiao et al., 2018; Wu et al., 2018). However, biochar modified with different iron materials has several immobilizations that are effective on arsenic.

7.3.14 VOLATILIZATION OF ARSENIC

Converting the Asi to methylated organic species like MMA and DMA and then to the gaseous trimethylarsine (TMA), is called volatilization. Rice plants that are transgenic which expresses *ArsM* produce advanced volatile arsenic, which helps in maintaining arsenic levels low in rice seeds. It also reduces organic arsenic species such as DMAV and MMAV in the roots and shoots of transgenic rice plants (Yang et al., 2015; Linquist et al., 2015; Hu et al., 2015; Verma et al., 2018).

7.3.15 WATER MANAGEMENT AND IRRIGATION PRACTICES

One of the most important regulators of the level and type of arsenic available to rice plants is the type of water management practices used in the field (Rahaman et al., 2011; Hua et al., 2013; Ma et al., 2014; Hu et al., 2015; Newbigging et al., 2015; Das et al., 2016; Norton et al., 2017; Wu et al., 2017). In recent years, the cultivation of rice under aerobic, intermittent flooding, and AWD (alternate wetting and drying condition) water management conditions has received considerable attention in terms of reducing the As content of rice while maintaining grain yields equivalent to those under conventional submerged conditions (Xu et al., 2008; Li et al., 2009; Deng et al., 2010; Somenahally et al., 2011; Talukder et al., 2011, 2012; Sarkar et al., 2012; Spanu et al., 2012; Ma et al., 2014; Moreno-Jim´enez et al., 2014; Hu et al., 2015; Linquist et al., 2015; Das et al., 2016; LaHue et al., 2016).

Sarkar et al. (2012) evaluated different irrigation practices such as continuous watering, intermittent watering, flooding, and aerobic to reduce the arsenic load on rice grains without affecting grain yield and further reported that soil arsenic loads were reduced in the order of continuous watering > intermittent accumulation > saturation > aerobic., while the arsenic concentration at the bottom of leaves and grains was obtained by intermittent accumulation. However, even unlike the continuous pond, the intermittent pond results in a slight decrease in yield in their experiment. Talukder et al. (2011) reported that aerobic rice cultivation increases grain productivity and decreases the concentration of arsenic it contains. A similar finding was made by Hu et al. (2015) for intermittent puddles, aerobic or AWD conditions as opposed to continuous flooding. Therefore, the amount of arsenic in rice and soils related to rice cultivation can be reduced through water-saving rice cultivation practices. Arsenic added by irrigation shows a more stable relationship with the arsenic status of different plant parts. Cultivating rice under deficit irrigation reduces arsenic pollution of the soil-roots-shoot-grain continuum (Stroud et al., 2011; Talukder et al., 2011; Sarkar et al., 2012; Newbigging et al., 2015).

Deficit irrigation can also reduce the availability of arsenic in the soil, regardless of how abundant arsenic is from contaminated irrigation water. Waterlogged conditions lead to a decrease in pH, an increase in pH in the As(III) segment in the solid phase of the soil, and this proportion can even increase up to 80% of the total arsenic in the soil (Yamaguchi et al., 2011; Pan et al., 2014; Syu et al., 2015; Rinklebe et al., 2016). The response-to-soil ratio of As(III) and As(V) increases due to flooding, while this ratio for As(III) is often higher than that of As(V), indicating that As(III) is more easily released of the soil solution than As(V) (Yamaguchi et al., 2011, 2014). Stroud et al. (2011); Talukder et al. (2012); and Hu et al. (2015) also observed greater mobilization of As(III) with a reduction of redox potential than that observed for As(V).

The decrease in arsenic content under aerobic and AWD conditions is usually linked to a relatively high soil redox potential, a low concentration of As(III) in the water of the soil pores, a decrease in the relative abundance of arsenic, bacteria that reduce Fe and S, an increase within the relative abundance of arsenic, Fe, and S-oxidizing bacteria and the formation of Fe plaques, which bind As(III) and As(V) to the surface of the rice root (Deng et al., 2010; Deng & Kim, 2016).

A major source of arsenic-contaminated irrigation water is a major source of arsenic in rice fields (Rahman et al., 2008; Adomako et al., 2009; Williams et al., 2009; Chatterjee et al., 2010; Arao et al., 2011a; Rahman & Hasegawa, 2011; Bhattacharya et al., 2013). As previously mentioned, numerous research efforts are underway to reduce arsenic intake in rice paddies by reducing the amount of water used for irrigation, for example by alternating humidification and drying or by using sprinkler irrigation methods. Additionally, the effectiveness of potentially arsenic-binding materials in reducing arsenic bioavailability was also tested. To reduce the concentration of arsenic in irrigation water before using it on rice fields, Polizzotto et al. (2015) demonstrated the effectiveness of improving irrigation networks with hessian buildings. Removal of all arsenic from the solution was improved in modified channels, with 7% of total arsenic removed compared to 3% in unmodified channels. The reason for this is attributed to an increase in the time interval and the efficiency of particle capture, whereas the removal via the formation of oxidizing elements does not seem to be significantly increased.

7.3.16 PHYTOREMEDIATION

The concentration of arsenic in irrigation water varies considerably over a range of 4–3,000 µg L–1 (Kramar et al., 2017). However, the application of Silicone, Phosphorous, iron, Sulfur organic matter, and biochar along with the action of soil microflora can help reduce the bioavailability of arsenic in soil. Therefore, growing crops that absorb and accrue an increased quantity of arsenic in their tissues and removing their biomass from field soil are effective approaches to reducing arsenic quantity, especially in soils that are already affected by high concentrations of arsenic. Pteris vittata L. is a well-known arsenic hyperaccumulator (Ma et al., 2001). Zhang et al. (2008) found an oversize diversity (50–280 mg As kg–1) in arsenic accumulation of 50 Azolla species when grown for 10 days in a solution medium with arsenic. Of the species tested, wild mosquito fern collected two folds of arsenic as Azolla filiculoides Lam. The prominent arsenic species in the two Azolla species, with the arsenic species (methylated) contributing less than 5% of the total arsenic concentration. It has also been reported that inoculation of field soil with Chlorella vulgaris Beijerinck and Nannochloropsis sp. growth caused by arsenic toxicity, restricted crop growth lowering protein concentration in rice (Upadhyay et al., 2016).

7.3.17 RICE POLISH AND VARIETY OF RICE

The outer layers of rice kernels give rise to rice luster through grinding and consist of a pericarp, seed coat, core, aleurone layer, germ, and part of the starchy endosperm subaleuronic layer with a high concentration of Si. The application of polished rice in the soil as a modification is an easier way to remove arsenic from the solution than other biosorbents (Ranjan et al., 2009). Adsorption of arsenic is often defined via

Langmuir, Freundlich, and DR (Dubinin-Radushkevich) isothermal models. The utmost absorbency calculated with the Langmuir model is 139 μg g–1 for As(III) and 147 μg g–1 for As(V) at the same temperature but with different pH. Therefore, rice glaze has the ability to absorb both As(III) and As(V).

In a study by Kumar et al. (2020), it was found that if the production of thiols and phytochelatins are stirred to an optimal level, it proves to be an aid in effectively tolerating arsenic toxicity in rice plants. They worked on two different genotypes of Oryza sativa-HAARG (with high arsenic content) and LAARG (with low arsenic content) and further found that the formation of a complex between phytochelatins and arsenic was comparatively more in HAARG than LAARG. Their study highlights a new concept that may be plants activated production, as well as, the consumption of thiols to counter arsenic toxicity. The phytotoxicity results in HAARG even after the accretion of a higher level of PCs is attributable to the disturbance caused by the high concentration of As(V) and As(III). On the contrary, LAARG displayed less up-regulation of Phytochelatins, however, it was passable to counter relatively low arsenic concentrations.

7.3.18 SULFUR IN SCALE-BACK ARSENIC

Non-proteins thiols with low molecular weight and a good quantity of sulfur are by-products of sulfur metabolism, and further play a vital role in the detoxification of arsenic. A study conducted by Dixit et al. (2015) reported that the appliance of sulfur can counterbalance the oxidative stress produced due to arsenic by decreasing the concentration of H2O2 and increasing antioxidant enzyme activities, furthermore, higher sulfur concentration leads to decreased OLsi1 and OLsi2 (transporter) level of expression which is attributable to the decreased accumulation of arsenic in a shoot which resulting in decreased quantity of arsenic in the food chain. The increased phytochelatins synthesis complexed arsenic in roots (Dixit et al., 2015). The research studies explored the role of sulfur in arsenic detoxification. It was reported that glycolytic enzymes play a vital role in amino acid biosynthesis, furthermore, the accrual of arsenic in shoots can be decreased by the supplementation of sulfur (Dixit et al., 2015) (Table 7.2).

TABLE 7.2 *Different Methods Reducing Arsenic Content in Rice Plants*

Sl. No.	Method	Mode of Application and Action	Arsenic Reduction (%) in Plant Parts	References
1.	Silicone	It can be applied directly in the soil and reduces the uptake of arsenic by increasing the amount of ferrihydrite in iron plaque.	25–50%, grain.	Seyffreth & Londrof (2012); Limmer et al. (2018)
2.	Nitrogen	It can be applied by means of hydroponic supplements and act by altering the structure of the root.	41%, seedling root	Srivastava et al. (2019)
3.	Sulfur	It can be applied by hydroponic means and improved metabolism including thiol and antioxidants.	Reduces arsenic	Zhang et al. (2011)
4.	Phosphorus	It can be directly applied to soil, which is naturally contaminated with high content of arsenic. It competes with the uptake mechanism of arsenate and improves redox homeostasis.	It reduces arsenic content by up to 10% in rice grains.	Talkuder et al. (2011)
5.	Biochar application	It can be applied directly. It reduced the uptake of arsenate and play a major role in iron plaque formation.	3, 6, and 14% reduction in arsenic content of rice grain when treated with 0.5, 1, and 2% manganese oxide-modified biochar.	Yu et al. (2017)
6.	Microbial method	Application of microbes including algae, fungi, etc., by means of hydroponic medium. It improved thiol metabolism including thiol synthesis and enzymes involved in the same.	80–90% reduction in arsenic by using AB_4O_2 and AB_4O_3; 45–52% reduction in accretion of arsenic in root and shoot.	Awasthi et al. (2018); Mallick et al. (2018)
7.	The alternate wetting-drying method in water management.	It involves the change in irrigation pattern, enhanced the aeration capability of the soil, and reduces the mobilization of inorganic soil.	60% reduction in rice grain.	Norton et al. (2017)
8.	Sprinkler irrigation	It involves the change in irrigation pattern and reduces the methylation process.	60–80% reduction of arsenic species in rice grain.	Moreno-Jimenez et al. (2014)
9.	Intermittent flooding	It involves the change in irrigation pattern and increased availability of silicone which in turn reduces the bioavailability of arsenic to plants.	40–60 and a reduction in the quantity of arsenic.	Majumadar et al. (2019); Srivastav et al. (2019)
10.	Raised bed cultivation	Furrow cultivation combines with Fe in soil and reduces the bioavailability of arsenic in the soil.	30% reduction in arsenic content	Duxbury & Panaulah (2007); Talkuder et al. (2011)

7.4 CONCLUSION

Issues which are related to the accretion of arsenic in rice plants are very wide. Accumulation of arsenic in rice depends on its bioavailability, and arsenic content in the soil. The accumulation is affected by factors like the types of soil, other factors like physicochemical, composition of phosphorus, iron, sulfur, and silicon in soil, presence of any other elements, organic, and inorganic matters, microbial populations, water management, etc. Approaches to diminish the accrual of arsenic in rice are adopting new method of agricultural practices and bioremediation methods. The species of arsenic that are methylated are less lethal than the Asi species. Other practices include microbial remediation which is resistant to arsenic, sowing of genetically modified rice plants (GM), using of rainwater for the harvested crop irrigation, adopting aerobic farming for paddy crops, and use of natural arsenic chelators are some of the effective ways which can control the arsenic contamination in rice grains. Molecular biology technologies have drawn attention to the various option for making rice grains free from arsenic accumulation. Changing and adopting new methods in/the practices of agriculture like using of sprinkler irrigation method stated to minimize arsenic accumulation in rice plants. Some soils are pre-contaminated with arsenic leading to a noxious amount of arsenic in rice and its parts. Rice is grown in flooded or waterlogged settings in most rice provinces. Though, submergence leads to an increase in the bioavailability of arsenic, with an upsurge in the fraction of highly toxic, abridged sorts of inorganic and organic arsenic. Hence, rice accrues further arsenic than other harvests, also the dynamics of arsenic in soil cultivating rice differs from that in upland harvesting systems and necessitates different agronomical intercessions which will be utilized in reducing its bioavailability. Consequently, an amalgamation of biochar, rice polish, Fe, P, or Si into the soil, aerobic or alternate wetting and drying water management, and selection of rice varieties that show improved rhizosphere aeration and formation of Fe plaques have shown promise for reducing arsenic availability in soil and uptake by rice plants.

KEYWORDS

- **agronomic measures**
- **arsenic**
- **bioavailability**
- **biochar**
- **heavy metal**
- **intermittent flooding**
- **rhizosphere aeration**
- **rice**
- **soil**

REFERENCES

Adomako, E. E., Solaiman, A. R. M., Williams, P. N., Deacon, C., Rahman, G. K. M. M., & Meharg, A. A., (2009). Enhanced transfer of arsenic to grain for Bangladesh grown rice compared to US and EU. *Environ. Int., 35*, 476–479.

Adriano, D. C., (2001). *Trace Elements in the Terrestrial Environments: Biogeochemistry Bioavailability, and Risks of Metals* (pp. 47–71). Springer: New York, USA.

Agrafioti, E., Kalderis, D., & Diamadopoulos, E., (2014). Ca and Fe modified biochars as adsorbents of arsenic and chromium in aqueous solutions. *J. Environ. Manage, 146*, 444–450.

Agrafioti, E., Kalderis, D., & Diamadopoulos, E., (2014). Ca and Fe modified biochars as adsorbents of arsenic and chromium in aqueous solutions. *J. Environ. Manage, 146*, 444–450.

Ahmed, Z. U., Panaullah, G. M., Gauch, H. G., McCouch, S. R., Ytagi, W., Kabir, M. S., & Duxbury, J. M., (2011). Genotype and environment effect rice (*Orza sativa* L.) grain arsenic concentration in Bangladesh. *Plant Soil, 338*, 367–382.

Al Agely, A., Sylvia, D. M., & Ma, L. Q., (2005). Mycorrhizae increase arsenic uptake by the hyperaccumulator Chinese brake fern (*Pteris vittata* L.). *J. Environ. Qual., 34*, 2181–2186.

Arao, T., Kawasaki, A., Baba, K., & Matsumoto, S., (2011a). Effects of arsenic compound amendment on arsenic speciation in rice grain. *Environ. Sci. Technol., 45*, 1291–1297.

Arao, T., Maejima, Y., & Baba, K., (2011b). Reduction in uptake by rice and soybean of aromatic arsenicals from diphenylarsinic acid contaminated soil amended with activated charcoal. *Environ Pollut., 159*, 2449–2453.

Arunakumara, K. K. I. U., Walpola, B. C., & Yoon, M. H., (2013). Current status of heavy metal contamination in Asia's rice lands. *Rev. Environ. Sci. Bio/Technol., 12*, 355–377.

Awasthi, G., Singh, A., Kumar, A., Singh, P., Kumar, S., Dwivedi, S., Trivedi, P., et al., (2015). Sulfur mediated reduction of arsenic toxicity involves efficient thiol metabolism and the antioxidant defense system in rice. *Journal of Hazardous Materials, 298*. 10.1016/j.jhazmat.2015.06.008.

Awasthi, S., Chauhan, R., Dwivedi, S., Srivastava, S., Srivastava, S., & Tripathi, R. D., (2018). A consortium of alga (*Chlorella vulgaris*) and bacterium (*Pseudomonas putida*) for amelioration of arsenic toxicity in rice: A promising and feasible approach. *Environ. Exp. Bot. 150*, 115–126. doi: 10.1016/j.envexpbot.2018.03.001.

Azam, S. M. G. G., Sarker, T. C., & Sabrina, N., (2016). Factors affecting the soil arsenic bioavailability, accumulation in rice and risk to human health: A review. *Toxicol. Mech. Method, 26,* 565–579.

Baig, J. A., Kazi, T. G., Shah, A. Q., Kandhro, G. A., Afridi, H. I., Khan, S., & Kolachi, N. F., (2010). Biosorption studies on powder of stem of *Acacia nilotica*: Removal of arsenic from surface water. *J. Hazard. Mater., 178*, 941–948.

Banerjee, M., Banerjee, N., Bhattacharjee, P., Mondal, D., Lythgoe, P. R., Martínez, M., Pan, J., et al., (2013). High arsenic in rice is associated with elevated genotoxic effects in humans. *Sci. Rep., 3*, 2195.

Bastías, J. M., & Beldarrain, T., (2016). Arsenic translocation in rice cultivation and its implication for human health. *Chil. J. Agric. Res., 76*, 114–122.

Batista, B. L., Bari˜ao, C. V., Souza, J. M. O., Paulelli, A. C. C., Rocha, B. A., De Oliveira, A. R. M., Segura, F. R., et al., (2016). A low-cost and environmentally-friendly potential procedure for inorganic As remediation based on the use of fungi isolated from rice rhizosphere. *J. Environ. Chem Eng., 4*, 891–898.

Beesley, L., & Marmiroli, M., (2011). The immobilization and retention of soluble arsenic, cadmium and zinc by biochar. *Environ. Pollut., 159*, 474–480.

Beesley, L., Moreno-Jim´enez, E., Gomez-Eyles, J. L., Harris, E., Robinson, B., & Sizmur, T., (2011). A review of biochars' potential role in the remediation, revegetation and restoration of contaminated soils. *Environ. Pollut., 159*, 3269–3282.

Bhattacharya, P., Jovanovic, D., & Polya, D., (2014). *Best Practice Guide on the Control of Arsenic in Drinking Water*. IWA Publishing: London, UK.

Bhattacharya, P., Mukherjee, A. B., Bundschuh, J., Zevenhoven, R., & Loeppert, R., (2007). *Arsenic in Soil and Groundwater Environment: Biogeochemical Interactions, Health Effects and Remediation* (Vol. 9). Elsevier: Amsterdam, The Netherlands.

Bhattacharya, P., Samal, A. C., Majumdar, J., & Santra, S. C., (2010). Accumulation of arsenic and its distribution in rice plant (*Oryza sativa* L.) in Gangetic West Bengal, India. *Paddy Water Environ., 8*, 63–70.

Bhattacharya, P., Samal, A. C., Majumdar, J., & Santra, S. C., (2010). Uptake of arsenic in rice plant varieties cultivated with arsenic rich groundwater. *Environ. Asia, 3*, 34–37.

Bhattacharya, P., Samal, A. C., Majumdar, J., Banerjee, S., & Santra, S. C., (2013). *In vitro* assessment on the impact of soil arsenic in the eight rice varieties of West Bengal, India. *J. Hazard Mater., 262*, 1091–1097.

Bhattacharya, S., Gupta, K., Debnath, S., Ghosh, U. C., Chattopadhyay, D. J., & Mukhopadhyay, A., (2012). Arsenic bioaccumulation in rice and edible plants and subsequent transmission through food chain in Bengal basin: A review of the perspectives for environmental health. *Toxicol. Environ. Chem., 94*, 429–441.

Bogdan, K., & Schenk, M. K., (2009). Evaluation of soil characteristics potentially affecting arsenic concentration in paddy rice (*Oryza sativa* L.). *Environ. Pollut., 157*, 2617–2621.
Bogdan, K., & Schenk, M. K., (2009). Evaluation of soil characteristics potentially affecting arsenic concentration in paddy rice (*Oryza sativa* L.). *Environ. Pollut., 157*, 2617–2621.
Cabrera, A., Cox, L., Spokas, K., Hermos´ın, M. C., Cornejo, J., & Koskinen, W. C., (2014). Influence of biochar amendments on the sorption-desorption of aminocyclopyrachlor, bentazone and pyraclostrobin pesticides to an agricultural soil. *Sci. Total Environ., 470, 471*, 438–443.
Cabrera, A., Cox, L., Spokas, K., Hermos´ın, M. C., Cornejo, J., & Koskinen, W. C., (2014). Influence of biochar amendments on the sorption-desorption of aminocyclopyrachlor, bentazone and pyraclostrobin pesticides to an agricultural soil. *Sci. Total Environ., 470, 471*, 438–443.
Campbell, J. A., Stark, J. H., & Carlton-Smith, C. H., (1985). *International Symposium on Heavy Metals in the Environment* (Vol. 1). CEP Consultants: Athens, Greece.

Carey, A. M., Norton, G. J., Deacon, C., Scheckel, K. G., Lombi, E., Punshon, T., Guerinot, M. L., Lanzirotti, A., Newville, M., Choi, Y., et al., (2011). Phloem transport of arsenic species from flag leaf to grain during grain filling. *New Phytol., 192*, 87–98.

Carey, A. M., Scheckel, K. G., Lombi, E., Newville, M., Choi, Y., Norton, G. J., Charnock, J. M., et al., (2010). Grain unloading of arsenic species in rice. *Plant Physiol., 152*, 309–319.

Chatterjee, D., Halder, D., Majumder, S., Biswas, A., Nath, B., Bhattacharya, P., Bhowmick, S., et al., (2010). Assessment of arsenic exposure from groundwater and rice in Bengal Delta Region, West Bengal, India. *Water Res., 44*, 5803–5812.

Chatterjee, S., Datta, S., Halder, M. P., Mitra, A., Veer, V., & Mukhopadhyay, S. K., (2013). Use of wetland plants in bioaccumulation of heavy metals. In: Gupta, D. K., (ed.), *Plant-Based Remediation Processes* (pp. 117–139). Springer: Berlin, Heidelberg, Germany.

Cope, C. O., Webster, D. S., & Sabatini, D. A., (2014). Arsenate adsorption onto iron oxide amended rice husk char. *Sci. Total Environ., 488, 489*, 554–561.

Dahal, B. M., Fuerhacker, M., Mentler, A., Karki, K. B., Shrestha, R. R., & Blum, W. E. H., (2008). Arsenic contamination of soils and agricultural plants through irrigation water in Nepal. *Environ. Pollut., 155*, 157–163.

Dang, F., Wang, W. X., Zhong, H., Wang, S. Q., Zhou, D. M., & Wang, Y., (2016). Effects of phosphate on trace element accumulation in rice (*Oryza sativa* L.): A 5-year phosphate application study. *J. Soil Sediment., 16*, 1440–1447.

Das, D. K., Sur, P., & Das, K., (2008). Mobilization of arsenic in soils and in rice (*Oryza sativa* L.) plants affected by organic matter and zinc application in irrigation water contaminated with arsenic. *Plant Soil Environ., 54*, 30–37.

Das, S., Chou, M. L., Jean, J. S., Liu, C. C., & Yang, H. J., (2016). Water management impacts on arsenic behavior and rhizosphere bacterial communities and activities in a rice agro-ecosystem. *Sci. Total Environ., 542*, 642–652.

Dasgupta, T., Hossain, S. A., Meharg, A. A., & Price, A. H., (2004). An arsenate tolerance gene on chromosome 6 of rice. *New Phytol., 163*, 45–49.

Delgado, A., & Go’mez, J. A., (2016). The soil. Physical, chemical and biological properties. In: Villalobos, F. J., & Fereres, E., (eds.), *Principles of Agronomy for Sustainable Agriculture* (pp. 15–26). Springer International Publishing AG: Cham, Switzerland.

Deng, B. L., & Kim, E. S., (2016). Co-adsorption of trichloroethylene and arsenate by iron-impregnated granular activated carbon. *Water Environ. Res., 88*, 394–402.

Deng, D., Wu, S. C., Wu, F. Y., Deng, H., & Wong, M. H., (2010). Effects of root anatomy and Fe plaque on arsenic uptake by rice seedlings grown in solution culture. *Environ. Pollut., 158*, 2589–2595.

Desplanques, V., Cary, L., Mouret, J. C., Trolard, F., Bourri´e, G., Grauby, O., & Meunier, J. D., (2006). Silicon transfers in a rice field in Camargue (France). *J. Geochem. Explor., 88*, 190–193.

Dixit, G., Singh, A. P., Kumar, A., Mishra, S., Dwivedi, S., Kumar, S., Trivedi, P. K., et al., (2016). Reduced arsenic accumulation in rice (*Oryza sativa* L.) shoot involves sulfur mediated improved thiol metabolism, antioxidant system and altered arsenic transporters. *Plant Physiology and Biochemistry: PPB, 99*, 86–96. https://doi.org/10.1016/j.plaphy.2015.11.005.

Dixit, G., Singh, A. P., Kumar, A., Singh, P. K., Kumar, S., Dwivedi, S., Trivedi, P. K., et al., (2015). Sulfur mediated reduction of arsenic toxicity involves efficient thiol metabolism and the antioxidant defense system in rice. *J. Hazard Mater., 298*, 241–251.

Dixit, G., Singh, A., Kumar, A., et al., (2015). Sulfur alleviates arsenic toxicity by reducing its accumulation and modulating proteome, amino acids and thiol metabolism in rice leaves. *Sci. Rep., 5*, 16205. https://doi.org/10.1038/srep16205.

Duan, G. L., Hu, Y., Liu, W. J., Kneer, R., Zhao, F. J., & Zhu, Y. G., (2011). Evidence for a role of phytochelatins in regulating arsenic accumulation in rice grain. *Environ. Exp. Bot., 71*, 416–421.

Duxbury, J. M., & Panaullah, G. M., (2007). *Remediation of Arsenic for Agriculture Sustainability* (pp. 1–28). Food Security Health Bangladesh. Rome: FAO.

Fahad, S., Hussain, S., Saud, S., Tanveer, M., Bajwa, A. A., Hassan, S., Shah, A. N., Ullah, A., Wu, C., Khan, F. A., et al., (2015). A biochar application protects rice pollen from high-temperature stress. *Plant Physiol. Biochem., 96*, 281–287.

Fan, J., Xia, X., Hu, Z., Ziadi, N., & Liu, C., (2013). Excessive sulfur supply reduces arsenic accumulation in brown rice. *Plant Soil Environ., 59*, 169–174.

Farooq, M. A., Islam, F., Ali, B., Najeeb, U., Mao, B., Gill, R. A., Yan, G., et al., (2016). Arsenic toxicity in plants: Cellular and molecular mechanisms of its transport and metabolism. *Environ. Exp. Bot., 132*, 42–52.

Farooq, S. H., Chandrasekharam, D., Berner, Z., Norra, S., & St¨uben, D., (2010). Influence of traditional agricultural practices on mobilization of arsenic from sediments to groundwater in Bengal delta. *Water Res., 44*, 5575–5588.

Farrow, E. M., Wang, J. M., Burken, J. G., Shi, H. L., Yan, W. G., Yang, J., Hua, B., & Deng, B. L., (2015). Reducing arsenic accumulation in rice grain through iron oxide amendment. *Ecotoxicol. Environ. Saf., 118*, 55–61.

Finnegan, P. M., & Chen, W. H., (2012). Arsenic toxicity: The effects on plant metabolism. *Front Physiol., 3*, 182.

Fleck, A. T., Mattusch, J., & Schenk, M. K., (2013). Silicon decreases the arsenic level in rice grain by limiting arsenite transport. *J. Plant Nutr. Soil Sci., 176*, 785–794.

Fu, Y. R., Chen, M., Bi, X., He, Y., Ren, L., Xiang, W., Qiao, S., et al., (2011). Occurrence of arsenic in brown rice and its relationship to soil properties from Hainan Island, China. *Environ. Pollut., 159*, 1757–1762.

Geng, C. N., Zhu, Y. G., Liu, W. J., & Smith, S. E., (2005). Arsenate uptake and translocation in seedlings of two genotypes of rice is affected by external phosphate concentrations. *Aquat. Bot., 83*, 321–331.

Gonzalez, P. S., Talano, M. A., Oller, A. L. W., Ibanez, S. G., Medina, M. I., & Agostini, E., (2014). Update on mechanisms involved in arsenic and chromium accumulation, translocation and homeostasis in plants. In: Gupta, D. K., & Chatterjee, S., (eds.), *Heavy Metal Remediation: Transport and Accumulation in Plants* (pp. 45–72). Nova Science Publishers: Hauppauge, NY, USA.

Guo, W., Hou, Y. L., Wang, S. G., & Zhu, Y. G., (2005). Effect of silicate on the growth and arsenate uptake by rice (*Oryza sativa* L.) seedlings in solution culture. *Plant Soil, 272*, 173–181.

Guo, W., Zhu, Y. G., Liu, W. J., Liang, Y. C., Geng, C. N., & Wang, S. G., (2007). Is the effect of silicon on rice uptake of arsenate (AsV) related to internal silicon concentrations, iron plaque and phosphate nutrition? *Environ. Pollut., 148*, 251–257.

Gupta, D. K., & Chatterjee, S., (2017). *Arsenic Contamination in the Environment: The Issues and Solutions.* Springer International Publishing AG: Cham, Switzerland.

Gupta, D. K., Srivastava, S., Huang, H. G., Romero-Puertas, M. C., & Sandalio, L. M., (2011). Arsenic tolerance and detoxification mechanisms in plants. In: Sherameti, I., & Varma, A., (eds.), *Detoxification of Heavy Metals* (pp. 169–180). Springer International Publishing AG: Cham, Switzerland.

Gupta, D. K., Tiwari, S., Razafindrabe, B. H. N., & Chatterjee, S., (2017). Arsenic contamination from historical aspects till present situation. In: Gupta, D. K., & Chatterjee, S., (eds.), *Arsenic Contamination in the Environment: The Issues and Solutions* (pp. 1–12). Springer International Publishing AG: Cham, Switzerland.

Hall, J. L., (2002). Cellular mechanisms for heavy metal detoxification and tolerance. *J. Exp. Bot., 53*, 1–11.

He, X. H., & Lilleskov, E., (2014). Arsenic uptake and phytoremediation potential by arbuscular mycorrhizal fungi. In: Solaiman, Z. M., Abbott, L. K., & Varma, A., (eds.) Mycorrhizal *Fungi: Use in Sustainable Agriculture and Land Restoration* (pp. 259–275). Springer, Berlin, Heidelberg.

Hernandez, L. E., Sobrino-Plata, J., Montero-Palmero, M. B., Carrasco-Gil, S., Flores-Caceres, M. L., Ortega-Villasante, C., & Escobar, C., (2015). Contribution of glutathione to the control of cellular redox homeostasis under toxic metal and metalloid stress. *J. Exp. Bot., 66*, 2901–2911.

Hu, P. J., Ouyang, Y. N., Wu, L. H., Shen, L. B., Luo, Y. M., & Christie, P., (2015). Effects of water management on arsenic and cadmium speciation and accumulation in an upland rice cultivar. *J. Environ. Sci., 27*, 225–231.

Hu, Z. Y., Zhu, Y. G., Li, M., Zhang, L. G., Cao, Z. H., & Smith, F. A., (2007). Sulfur (S)-induced enhancement of iron plaque formation in the rhizosphere reduces arsenic accumulation in rice (*Oryza sativa* L.) seedlings. *Environ. Pollut., 147*, 387–393.

Hua, B., Yan, W. G., & Yang, J., (2013). Response of rice genotype to straight head disease as influenced by arsenic level and water management practices in soil. *Sci. Total Environ. 442*, 432–436.

Huang, J. H., Fecher, P., Ilgen, G., Hu, K. N., & Yang, J., (2012). Speciation of arsenite and arsenate in rice grain-verification of nitric acid based extraction method and mass sample survey. *Food Chem., 130*, 453–459.

Islam, F. S., Gault, A. G., Boothman, C., Polya, D. A., Charnock, J. M., Chatterjee, D., & Lloyd, J. R., (2004). Role of metal-reducing bacteria in arsenic release from Bengal delta sediment. *Nature, 430*, 68–71.

Jia, Y., Huang, H., Zhong, M., Wang, F. H., Zhang, L. M., & Zhu, Y. G., (2013). Microbial arsenic methylation in soil and rice rhizosphere. *Environ. Sci. Technol., 47*, 3141–3148.

Jia, Y., Sun, G. X., Huang, H., & Zhu, Y. G., (2013). Biogas slurry application elevated arsenic accumulation in rice plant through increased arsenic release and methylation in paddy soil. *Plant Soil., 365*, 387–396.

Kamiya, T., Islam, M. R., Duan, G., Uraguchi, S., & Fujiwara, T., (2013). Phosphate deficiency signaling pathway is a target of arsenate and phosphate transporter OsPT1 is involved in As accumulation in shoots of rice. *Soil Sci. Plant Nutr., 59*, 580–590.

Kar, S., Das, S., Jean, J. S., Chakraborty, S., & Liu, C. C., (2013). Arsenic in the water-soil-plant system and the potential health risks in the coastal part of Chianan Plain, Southwestern Taiwan. *J. Asian Earth Sci., 77*, 295–302.

Kraemer, S., (2004). Iron oxide dissolution and solubility in the presence of siderophores. *Aquat. Sci. Res., 66*, 3–18.

Kramar, U., Norra, S., Berner, Z., Kiczka, M., & Chandrasekharam, D., (2017). On the distribution and speciation of arsenic in the soil-plant-system of a rice field in West-Bengal, India: A μsynchrotron techniques based case study. *Appl. Geochem., 77*, 4–14.

Kumar, A., Singh, P., Srivastava, S., Dwivedi, S., Tripathi, R., Awasthi, G., Gupta, K., & Ansari, M., (2020). A comparative study on the effect of arsenic on thiolic ligands and phytochelatins in contrasting arsenic accumulating rice genotypes. *International Journal of Plant and Environment, 6*(02), 110–117. https://doi.org/10.18811/ijpen.v6i02.03.

LaHue, G. T., Chaney, R. L., Adviento-Borbe, M. A., & Linquist, B. A., (2016). Alternate wetting and drying in high yielding direct seeded rice systems accomplishes multiple environmental and agronomic objectives. *Agric. Ecosyst. Environ., 229*, 30–39.

Lauren, J. G., & Duxbury, J. M., (2005). Management strategies to reduce arsenic uptake by rice. In: *Proceedings of the International Symposium on Behavior of Arsenic in Aquifers, Soils and Plants: Implications for Management*. Dhaka, Bangladesh. Centro Internacional de Mejoramiento de Maíz y Trigo and the U.S. Geological Survey: Reston, VA, USA.

Lehmann, J., (2007). A handful of carbon. *Nature, 447*, 143, 144.

Li, R. Y., Ago, Y., Liu, W. J., Mitani, N., Feldmann, J., McGrath, S. P., Ma, J. F., & Zhao, F. J., (2009). The rice aquaporin Lsi1 mediates uptake of methylated arsenic species. *Plant Physiol., 150*, 2071–2080.

Li, R. Y., Stroud, J. L., Ma, J. F., Mcgrath, S. P., & Zhao, F. J., (2009). Mitigation of arsenic accumulation in rice with water management and silicon fertilization. *Environ. Sci. Technol., 43*, 3778–3783.

Linquist, B. A., Anders, M. M., Adviento-Borbe, M. A. A., Chaney, R. L., Nalley, L. L., Da Rosa, E. F. F., & Van, K. C., (2015). Reducing greenhouse gas emissions, water use, and grain arsenic levels in rice systems. *Global Change Biol., 21*, 407–417.

Linquist, B. A., Anders, M. M., Adviento-Borbe, M. A. A., Chaney, R. L., Nalley, L. L., Da Rosa, E. F. F., & Van, K. C., (2015). Reducing greenhouse gas emissions, water use, and grain arsenic levels in rice systems. *Global Change Biol., 21*, 407–417.

Liu, W. J., McGrath, S. P., & Zhao, F. J., (2014). Silicon has opposite effects on the accumulation of inorganic and methylated arsenic species in rice. *Plant Soil, 376*, 423–431.

Liu, W. J., Zhu, Y. G., Smith, F. A., & Smith, S. E., (2004). Do iron plaque and genotypes affect arsenate uptake and translocation by rice seedlings (*Oryza sativa* L.) grown in solution culture? *J. Exp. Bot., 55*, 1707–1713.

Liu, W. J., Zhu, Y. G., Smith, F. A., & Smith, S. E., (2004). Do iron plaque and genotypes affect arsenate uptake and translocation by rice seedlings (*Oryza sativa* L.) grown in solution culture? *J. Exp. Bot., 55*, 1707–1713.

Lomax, C., Liu, W. J., Wu, L. Y., Xue, K., Xiong, J. B., Zhou, J. Z., McGrath, S. P., et al., (2012). Methylated arsenic species in plants originate from soil microorganisms. *New Phytol., 193*, 665–672.

Ma, J. F., Tamai, K., Yamaji, N., Mitani, N., Konishi, S., Katsuhara, M., Ishiguro, M., et al., (2006). A silicon transporter in rice. *Nature, 440*, 688–691.

Ma, J. F., Yamaji, N., Mitani, N., Tamai, K., Konishi, S., Fujiwara, T., Katsuhara, M., & Yano, M., (2007). An efflux transporter of silicon in rice. *Nature, 448*, 209–212.

Ma, J. F., Yamaji, N., Mitani, N., Xu, X. Y., Su, Y. H., McGrath, S. P., & Zhao, F. J., (2008). Transporters of arsenite in rice and their role in arsenic accumulation in rice grain. *Proc. Natl. Acad. Sci. USA, 105*, 9931–9935.

Ma, J. F., Yamaji, N., Mitani, N., Xu, X. Y., Su, Y. H., McGrath, S. P., & Zhao, F. J., (2008). Transporters of arsenite in rice and their role in arsenic accumulation in rice grain. *Proc. Natl. Acad. Sci. USA., 105*, 9931–9935.

Ma, L. Q., Komar, K. M., Tu, C., Zhang, W. H., Cai, Y., & Kennelley, E. D., (2001). A fern that hyperaccumulates arsenic. *Nature, 409*, 579.

Ma, R., Shen, J. L., Wu, J. S., Tang, Z., Shen, Q. R., & Zhao, F. J., (2014). Impact of agronomic practices on arsenic accumulation and speciation in rice grain. *Environ. Pollut. 194,* 217–223.

Majumdar, A., Upadhyay, M. K., Kumar, J. S., Barla, A., Srivastava, S., Jaiswal, M. K., et al., (2019). Ultra-structure alteration via enhanced silicon uptake in arsenic stressed rice cultivars under intermittent irrigation practices in Bengal delta basin. *Ecotoxicol. Environ. Saf., 180*, 770–779. doi: 10.1016/j.ecoenv.2019.05.028\.

Malasarn, D., Saltikov, C. W., Campbell, K. M., Santini, J. M., Hering, J. G., & Newman, D. K., (2004). arrA is a reliable marker for As(V) respiration. *Science, 306*, 455.

Mallick, I., Bhattacharyya, C., Mukherji, S., Dey, D., Sarkar, S. C., Mukhopadhyay, U. K., et al., (2018). Effective rhizoinoculation and biofilm formation by arsenic immobilizing halophilic plant growth promoting bacteria (PGPB) isolated from mangrove rhizosphere: A step towards arsenic rhizoremediation. *Sci. Total Environ., 610*, 1239–1250. doi: 10.1016/j.scitotenv.2017.07.234.

Marin, A. R., Masscheleyn, P. H., & Patrick, W. H., (1992). The influence of chemical form and concentration of arsenic on rice growth and tissue arsenic concentration. *Plant Soil, 139*, 175–183.

Martinez, V. D., Vucic, E. A., Becker-Santos, D. D., Gil, L., & Lam, W. L., (2011). Arsenic exposure and the induction of human cancers. *J. Toxicol., 2011*, 431287.

Meharg, A. A., & Rahman, M. D. M., (2003). Arsenic contamination of Bangladesh paddy field soils: Implications for rice contribution to arsenic consumption. *Environ. Sci. Technol., 37*, 229–234.

Meharg, A. A., & Zhao, F. J., (2012). *Arsenic & Rice*. Springer International Publishing AG: Cham, Switzerland.

Meharg, C., & Meharg, A. A., (2015). Silicon, the silver bullet for mitigating biotic and abiotic stress, and improving grain quality, in rice? *Environ. Exp. Bot., 120*, 8–17.

Mitra, A., Chatterjee, S., & Gupta, D. K., (2017). Uptake, transport, and remediation of arsenic by algae and higher plants. In: Gupta, D. K., & Chatterjee, S., (eds.), *Arsenic Contamination in the Environment* (pp. 145–169). Springer International Publishing AG: Cham, Switzerland.

Mitra, A., Chatterjee, S., Datta, S., Sharma, S., Veer, V., Razafindrabe, B. H. M., Walther, C., & Gupta, D. K., (2014). Mechanism of metal transporter in plants. In: Gupta, D. K., & Chatterjee, S., (eds.), *Heavy Metal Remediation: Transport and Accumulation in Plants* (pp. 1–27). Nova Science Publishers: Hauppauge, NY, USA.

Moreno-Jim´enez, E., Meharg, A. A., Smolders, E., Manzano, R., Becerra, D., S´anchez-Llerena, J., Albarr´an, A., & L´opez-Pi˜nero, A., (2014). Sprinkler irrigation of rice fields reduces grain arsenic but enhances cadmium. *Sci. Total Environ., 485, 486*, 468–473.

Newbigging, A. M., Paliwoda, R. E., & Le, X. C., (2015). Rice: Reducing arsenic content by controlling water irrigation. *J. Environ. Sci., 30*, 129–131.

Norton, G. J., Deacon, C. M., Xiong, L., Huang, S., Meharg, A. A., & Price, A. H., (2010). Genetic mapping of the rice ionome in leaves and grain: Identification of QTLs for 17 elements including arsenic, cadmium, iron and selenium. *Plant Soil, 329*, 139–153.

Norton, G. J., Duan, G., Dasgupta, T., Islam, M. R., Lei, M., Zhu, Y., Deacon, C. M., Moran, A. C., Islam, S., Zhao, F. J., et al., (2009). Environmental and genetic control of arsenic accumulation and speciation in rice grain: Comparing a range of common cultivars grown in contaminated sites across Bangladesh, China and India. *Environ. Sci. Technol., 43*, 8381–8386.

Norton, G. J., Islam, M. R., Deacon, C. M., Zhao, F. J., Stroud, J. L., Mcgrath, S. P., Islam, S., Jahiruddin, M., Feldmann, J., Price, A. H., et al., (2009). Identification of low inorganic and total grain arsenic rice cultivars from Bangladesh. *Environ. Sci. Technol., 43*, 6070–6075.

Norton, G. J., Shafaei, M., Travis, A. J., Deacon, C. M., Danku, J., Pond, D., Cochrane, N., et al., (2017). Impact of alternate wetting and drying on rice physiology, grain production, and grain quality. *Field Crop Res., 205*, 1–13.

Pan, W. S., Wu, C., Xue, S. G., & Hartley, W., (2014). Arsenic dynamics in the rhizosphere and its sequestration on rice roots as affected by root oxidation. *J. Environ. Sci., 26*, 892–899.

Paszkowski, U., Kroken, S., Roux, C., & Briggs, S. P., (2002). Rice phosphate transporters include an evolutionarily divergent gene specifically activated in arbuscular mycorrhizal symbiosis. *Proc. Natl. Acad. Sci. USA, 99*, 13324–13329.

Pikaray, S., Banerjee, S., & Mukherji, S., (2005). Sorption of arsenic onto Vindhyan shales: Role of pyrite and organic carbon. *Curr. Sci., 88*, 1580–1585.

Pinson, S. R. M., Tarpley, L., Yan, W., Yeater, K., Lahner, B., Yakubova, E., Huang, X. Y., et al., (2015). Worldwide genetic diversity for mineral element concentrations in rice grain. *Crop Sci., 55*, 294–311.

Polizzotto, M. L., Birgand, F., Badruzzaman, A. B. M., & Ali, M. A., (2015). Amending irrigation channels with jute-mesh structures to decrease arsenic loading to rice fields in Bangladesh. *Ecol. Eng., 74*, 101–106.

Punshon, T., Jackson, B. P., Meharg, A. A., Warczack, T., Scheckel, K., & Guerinot, M. L., (2017). Understanding arsenic dynamics in agronomic systems to predict and prevent uptake by crop plants. *Sci. Total Environ., 581, 582*, 209–220.

Qiao, J. T., Liu, T. X., Wang, X. Q., Li, F. B., Lv, Y. H., Cui, J. H., Zeng, X. D., et al., (2018). Simultaneous alleviation of cadmium and arsenic accumulation in rice by applying zero-valent iron and biochar to contaminated paddy soils. *Chemosphere, 195*, 260–271. https://doi. org/10.1016/j.chemosphere.2017.12.081.

Qin, J., Rosen, B. P., Zhang, Y., Wang, G. J., Franke, S., & Rensing, C., (2006). Arsenic detoxification and evolution of trimethylarsine gas by a microbial arsenite S-adenosylmethionine methyltransferase. *Proc. Natl. Acad. Sci. USA., 103*, 2075–2080.

Quazi, S., Datta, R., & Sarkar, D., (2011). Effect of soil types and forms of arsenical pesticide on rice growth and development. *Int. J. Environ. Sci. Technol., 8*, 450–460.

Raab, A., Williams, P. N., Meharg, A., & Feldmann, J., (2007). Uptake and translocation of inorganic and methylated arsenic species by plants. *Environ. Chem., 4*, 197–203.

Rahaman, S., Sinha, A. C., & Mukhopadhyay, D., (2011). Effect of water regimes and organic matters on transport of arsenic in summer rice (*Oryza sativa* L.). *J. Environ. Sci., 23*, 633–639.

Rahaman, S., Sinha, A. C., & Mukhopadhyay, D., (2011). Effect of water regimes and organic matters on transport of arsenic in summer rice (*Oryza sativa* L.). *J. Environ. Sci., 23*, 633–639.

Rahman, M. A., & Hasegawa, H., (2011). High levels of inorganic arsenic in rice in areas where arsenic-contaminated water is used for irrigation and cooking. *Sci. Total Environ., 409*, 4645–4655.

Rahman, M. A., Hasegawa, H., Rahman, M. M., Miah, M. A. M., & Tasmin, A., (2008). Straight head disease of rice (*Oryza sativa* L.) induced by arsenic toxicity. *Environ. Exp. Bot., 62*, 54–59.

Ranjan, D., Talat, M., & Hasan, S. H., (2009). Biosorption of arsenic from aqueous solution using agricultural residue 'rice polish.' *J. Hazard Mater., 166*, 1050–1059.

Rinklebe, J., Shaheen, S. M., & Yu, K. W., (2016). Release of As, Ba, Cd, Cu, Pb, and Sr under pre-definite redox conditions in different rice paddy soils originating from the U.S.A. and Asia. *Geoderma., 270*, 21–32.

Sadiq, M., (1997). Arsenic chemistry in soils: An overview of thermodynamic predictions and field observations. *Water Air Soil Pollut., 93*, 117–136.

Saha, G. C., & Ali, M. A., (2007). Dynamics of arsenic in agricultural soils irrigated with arsenic contaminated groundwater in Bangladesh. *Sci. Total Environ., 379*, 180–189.

Sahoo, P. K., & Kim, K., (2013). A review of the arsenic concentration in paddy rice from the perspective of geoscience. *Geosci. J., 17*, 107–122.

Samsuri, A. W., Sadegh-Zadeh, F., & Seh-Bardan, B. J., (2013). Adsorption of As(III) and As(V) by Fe coated biochars and biochars produced from empty fruit bunch and rice husk. *J. Environ. Chem. Eng., 1*, 981–988.

Sanglard, L. M. V. P., Detmann, K. C., Martins, S. C. V., Teixeira, R. A., Pereira, L. F., Sanglard, M. L., Fernie, A. R., et al., (2016). The role of silicon in metabolic acclimation of rice plants challenged with arsenic. *Environ. Exp. Bot., 123*, 22–36.

Sarkar, S., Basu, B., Kundu, C. K., & Patra, P. K., (2012). Deficit irrigation: An option to mitigate arsenic load of rice grain in West Bengal, India. *Agric. Ecosyst. Environ., 146*, 147–152.

Seyfferth, A. L., & Fendorf, S., (2012). Silicate mineral impacts on the uptake and storage of arsenic and plant nutrients in rice (*Oryza sativa* L.). *Environ. Sci. Technol., 46*, 13176–13183.

Seyfferth, A. L., Morris, A. H., Gill, R., Kearns, K. A., Mann, J. N., Paukett, M., & Leskanic, C., (2016). Soil incorporation of silica-rich rice husk decreases inorganic arsenic in rice grain. *J. Agric. Food Chem., 64*, 3760–3766.

Shi, S., Wang, T., Chen, Z., Tang, Z., Wu, Z., & Salt, D. E., (2016). OsHAC1; 1 and OsHAC1; 2 function as arsenate reductases and regulate arsenic accumulation. *Plant Physiol., 172*, 1708–1719.

Shrivastava, A., Barla, A., Majumdar, A., Singh, S., & Bose, S., (2019). Arsenic mitigation in rice grain loading via alternative irrigation by proposed water management practices. *Chemosphere, 238*, 124988. doi: 10.1016/j.chemosphere.2019.124988.

Signes-Pastor, A., Burló, F., Mitra, K., & Carbonell-Barrachina, A. A., (2007). Arsenic biogeochemistry as affected by phosphorus fertilizer addition, redox potential and pH in a West Bengal (India) soil. *Geoderma, 137*, 504–510.

Singh, H. P., Kaur, S., Batish, D. R., Sharma, V. P., Sharma, N., & Kohli, R. K., (2009). Nitric oxide alleviates arsenic toxicity by reducing oxidative damage in the roots of *Oryza sativa* (rice). *Nitric Oxide., 20*, 289–297.

Singh, N., Marwa, N., Mishra, S. K., Mishra, J., Verma, P. C., Rathaur, S., & Singh, N., (2016a). *Brevundimonas diminuta* mediated alleviation of arsenic toxicity and plant growth promotion in *Oryza sativa* L. *Ecotoxicol. Environ. Saf., 125*, 25–34.

Somenahally, A. C., Hollister, E. B., Loeppert, R. H., Yan, W. G., & Gentry, T. J., (2011). Microbial communities in rice rhizosphere altered by intermittent and continuous flooding in fields with long-term arsenic application. *Soil Biol. Biochem., 43*, 1220–1228.

Song, W. Y., Yamaki, T., Yamaji, N., Ko, D., Jung, K. H., Fujii-Kashino, M., An, G., et al., (2014). A rice ABC transporter, OsABCC1, reduces arsenic accumulation in the grain. *Proc. Natl. Acad. Sci. USA, 111*, 699–704.

Song, W. Y., Yamaki, T., Yamaji, N., Ko, D., Jung, K. H., Fujii-Kashino, M., An, G., et al., (2014). A rice ABC transporter, OsABCC1, reduces arsenic accumulation in the grain. *Proc. Natl. Acad. Sci. USA, 111*, 699–704.

Spanu, A., Daga, L., Orlandoni, A. M., & Sanna, G., (2012). The role of irrigation techniques in arsenic bioaccumulation in rice (*Oryza sativa* L.). *Environ. Sci. Technol., 46*, 8333–8340.

Stroud, J. L., Norton, G. J., Islam, M. R., Dasgupta, T., White, R. P., Price, A. H., Meharg, A. A., et al., (2011). The dynamics of arsenic in four paddy fields in the Bengal delta. *Environ. Pollut., 159*, 947–953.

Suriyagoda, L. D. B., Dittert, K., & Lambers, H., (2018). Mechanism of arsenic uptake, translocation and plant resistance to accumulate arsenic in rice grains. *Agric. Ecosyst. Environ., 253*, 23–37.

Suriyagoda, L., Dittert, K., & Lambers, H., (2018). Arsenic in rice soils and potential agronomic mitigation strategies to reduce arsenic bioavailability: A review. *Pedosphere, 28*, 363–382. 10.1016/S1002-0160(18)60026-8.

Syu, C. H., Huang, C. C., Jiang, P. Y., Lee, C. H., & Lee, D. Y., (2015). Arsenic accumulation and speciation in rice grains influenced by arsenic phytotoxicity and rice genotypes grown in arsenic-elevated paddy soils. *J. Hazard. Mater., 286*, 179–186.

Syu, C. H., Huang, C. C., Jiang, P. Y., Lee, C. H., & Lee, D. Y., (2015). Arsenic accumulation and speciation in rice grains influenced by arsenic phytotoxicity and rice genotypes grown in arsenic elevated paddy soils. *J. Hazard Mater., 286*, 179–186.

Takahashi, Y., Minamikawa, R., Hattori, K. H., Kurishima, K., Kihou, N., & Yuita, K., (2004). Arsenic behavior in paddy fields during the cycle of flooded and non-flooded periods. *Environ. Sci. Technol., 38*, 1038–1044.

Talukder, A. S. M. H. M., Meisner, C. A., Sarkar, M. A. R., & Islam, M. S., (2011). Effect of water management, tillage options and phosphorus status on arsenic uptake in rice. *Ecotoxicol. Environ. Saf., 74*, 834–839.

Talukder, A. S. M. H. M., Meisner, C. A., Sarkar, M. A. R., Islam, M. S., Sayre, K. D., Duxbury, J. M., & Lauren, J. G., (2012). Effect of water management, arsenic and phosphorus levels on rice in a high-arsenic soil-water system: II. Arsenic uptake. *Ecotoxicol. Environ. Saf., 80,* 145–151.

Tang, J. C., Zhu, W. Y., Kookana, R., & Katayama, A., (2013). Characteristics of biochar and its application in remediation of contaminated soil. *J. Biosci. Bioeng., 116*, 653–659.

Trotta, A., Falaschi, P., Cornara, L., Minganti, V., Fusconi, A., Drava, G., & Berta, G., (2006). Arbuscular mycorrhizae increase the arsenic translocation factor in the As hyperaccumulating fern *Pteris vittata* L. *Chemosphere, 65*, 74–81.

Upadhyay, A. K., Singh, N. K., Singh, R., & Rai, U. N., (2016). Amelioration of arsenic toxicity in rice: Comparative effect of inoculation of *Chlorella vulgaris* and nannochloropsis sp. on growth, biochemical changes and arsenic uptake. *Ecotoxicol. Environ. Saf., 124*, 68–73.

Upadhyay, M. K., Majumdar, A., Suresh, K. J., & Srivastava, S., (2020). Arsenic in rice agro-ecosystem: Solutions for safe and sustainable rice production. *Frontiers in Sustainable Food Systems, 4*. doi=10.3389/fsufs.2020.00053.

Verma, P. K., Verma, S., Pande, V., Mallick, S., Deo, T. R., Dhankher, O. P., & Chakrabarty, D., (2016a). Overexpression of rice glutaredoxin OsGrx_C7 and OsGrx_C2. 1 reduces intracellular arsenic accumulation and increases tolerance in *Arabidopsis thaliana. Frontiers in Plant Science*, *7*, 740.

Verma, P. K., Verma, S., Tripathi, R. D., & Chakrabarty, D., (2020). A rice glutaredoxin regulates the expression of aquaporin genes and modulate root responses to provide arsenic tolerance. *Ecotoxicology and Environmental Safety, 195*, 110471.

Verma, S., Verma, P. K., Meher, A. K., Bansiwal, A. K., Tripathi, R. D., & Chakrabarty, D., (2018). A novel fungal arsenic methyltransferase, WaarsM reduces grain arsenic accumulation in transgenic rice (*Oryza sativa* L.). *Journal of Hazardous Materials, 344*, 626–634.

Verma, S., Verma, P. K., Meher, A. K., Dwivedi, S., Bansiwal, A. K., Pande, V., Srivastava, P. K., et al., (2016b). A novel arsenic methyltransferase gene of *Westerdykella aurantiaca* isolated from arsenic contaminated soil: Phylogenetic, physiological, and biochemical studies and its role in arsenic bioremediation. *Metallomics, 8*, 344–353.

Wang, P., Zhang, W., Mao, C., Xu, G., & Zhao, F. J., (2016). The role of OsPT8 in arsenate uptake and varietal difference in arsenate tolerance in rice. *J. Exp. Bot., 67*, 6051–6059.

Wang, S. L., & Mulligan, C. N., (2006). Effect of natural organic matter on arsenic release from soils and sediments into groundwater. *Environ. Geochem Health., 28*, 197–214.

Wani, P. A., Khan, M. S., & Zaidi, A., (2007). Effect of metal tolerant plant growth promoting *Bradyrhizobium* sp. (Vigna) on growth, symbiosis, seed yield and metal uptake by greengram plants. *Chemosphere, 70*, 36–45.

Wenzel, W. W., (2009). Rhizosphere processes and management in plant assisted bioremediation (phytoremediation) of soils. *Plant Soil, 321*, 385–408.

Williams, P. N., Lei, M., Sun, G. X., Huang, Q., Lu, Y., Deacon, C., Meharg, A. A., & Zhu, Y. G., (2009). Occurrence and partitioning of cadmium, arsenic and lead in mine impacted paddy rice: Hunan, China. *Environ. Sci. Technol., 43*, 637–642.

Williams, P. N., Villada, A., Deacon, C., Raab, A., Figuerola, J., Green, A. J., Feldmann, J., & Meharg, A. A., (2007). Greatly enhanced arsenic shoot assimilation in rice leads to elevated grain levels compared to wheat and barley. *Environ. Sci. Technol., 41*, 6854–6859.

Williams, P. N., Zhang, H., Davison, W., Mehrag, A. A., Hossain, M., Norton, G., Brammer, H., & Islam, M. R., (2011). Organic matter solid phase interactions are critical for predicting arsenic release and plant uptake in Bangladesh paddy soils. *Environ. Sci. Technol., 45*, 6080–6087.

Woolf, D., Amonette, J. E., Street-Perrott, F. A., Lehmann, J., & Joseph, S., (2010). Sustainable biochar to mitigate global climate change. *Nat. Commun., 1*, 56.

Wu, C., Cui, M., Xue, S., Li, W., Huang, L., Jiang, X., & Qian, Z., (2018). Remediation of arsenic-contaminated paddy soil by iron-modified biochar. *Environ. Sci. Pollut. Res., 25*, 20792–20801. https://doi.org/10. 1007/s11356-018-2268-8.

Wu, C., Huang, L., Xue, S. G., Pan, W. S., Zou, Q., Hartley, W., & Wong, M. H., (2017). Oxic and anoxic conditions affect arsenic (As) accumulation and arsenite transporter expression in rice. *Chemosphere, 168*, 969–975.

Wu, Z., Ren, H., McGrath, S. P., Wu, P., & Zhao, F. J., (2011). Investigating the contribution of the phosphate transport pathway to arsenic accumulation in rice. *Plant Physiol., 157*, 498–508.

Xiao, K. Q., Li, L. G., Ma, L. P., Zhang, S. Y., Bao, P., Zhang, T., & Zhu, Y. G., (2016). Metagenomic analysis revealed highly diverse microbial arsenic metabolism genes in paddy soils with low-arsenic contents. *Environ. Pollut., 211*, 1–8.

Xu, J., Shi, S., Wang, L., Tang, Z., Lv, T., Zhu, X., Ding, X., et al., (2017). OsHAC4 is critical for arsenate tolerance and regulates arsenic accumulation in rice. *New Phytol., 215*, 1090–1101.

Xu, X. Y., McGrath, S. P., Meharg, A. A., & Zhao, F. J., (2008). Growing rice aerobically markedly decreases arsenic accumulation. *Environ. Sci. Technol., 42*, 5574–5579.

Yamaguchi, N., Nakamura, T., Dong, D., Takahashi, Y., Amachi, S., & Makino, T., (2011). Arsenic release from flooded paddy soils is influenced by speciation, Eh, pH, and iron dissolution. *Chemosphere., 83*, 925–932.

Yang, G. D., Xie, W. Y., Zhu, X., Huang, Y., Yang, X. J., Qiu, Z. Q., Lv, Z. M., et al., (2015). Effect of arsenite-oxidizing bacterium B. laterosporus on arsenite toxicity and arsenic translocation in rice seedlings. *Ecotoxicol. Environ. Saf., 120*, 7–12.

Yang, G. D., Xie, W. Y., Zhu, X., Huang, Y., Yang, X. J., Qiu, Z. Q., Lv, Z. M., et al., (2015). Effect of arsenite-oxidizing bacterium B. laterosporus on arsenite toxicity and arsenic translocation in rice seedlings. *Ecotoxicol. Environ. Saf., 120*, 7–12.

Yao, Y., Zhou, H., Yan, X. L., Yang, X., Huang, K. W., Liu, J., Li, L. J., et al., (2021). The Fe_3O_4-modified biochar reduces arsenic availability in soil and arsenic accumulation in indica rice (*Oryza sativa* L.). *Environmental Science and Pollution Research International, 28*(14), 18050–18061. https://doi.org/10.1007/s11356-020-11812-x.

Ye, X. X., Sun, B., & Yin, Y. L., (2012). Variation of As concentration between soil types and rice genotypes and the selection of cultivars for reducing As in the diet. *Chemosphere, 87*, 384–389.

Yu, H. Y., Ding, X. D., Li, F. B., Wang, X. Q., Zhang, S. R., Yi, J. C., Liu, C. P., et al., (2016). The availabilities of arsenic and cadmium in rice paddy fields from a mining area: The role of soil extractable and plant silicon. *Environ. Pollut., 215*, 258–265.

Yu, Z., Qiu, W., Wang, F., Lei, M., Wang, D., & Song, Z., (2017). Effects of manganese oxide-modified biochar composites on arsenic speciation and accumulation in an indica rice (*Oryza sativa* L.) cultivar. *Chemosphere, 168*, 341–349. doi: 10.1016/j.chemosphere.2016.10.069.

Zhang, J., & Duan, G. L., (2008). Genotypic difference in arsenic and cadmium accumulation by rice seedlings grown in hydroponics. *J. Plant Nutr., 31*, 2168–2182.

Zhang, J., Zhao, Q. Z., Duan, G. L., & Huang, Y. C., (2011). Influence of sulfur on arsenic accumulation and metabolism in rice seedlings. *Environ. Exp. Bot., 72*, 34–40.

Zhang, X., Lin, A. J., Zhao, F. J., Xu, G. Z., Duan, G. L., & Zhu, Y. G., (2008). Arsenic accumulation by the aquatic fern Azolla: Comparison of arsenate uptake, speciation and efflux by A. caroliniana and A. filiculoides. *Environ. Pollut., 156*, 1149–1155.

Zhao, F. J., Ma, J. F., Meharg, A. A., & Mc Grath, S. P., (2008). Arsenic uptake and metabolism in plants. *New Phytol., 181*, 777–794.

Zhao, F. J., McGrath, S. P., & Meharg, A. A., (2010). Arsenic as a food chain contaminant: Mechanisms of plant uptake and metabolism and mitigation strategies. *Annu. Rev. Plant Biol., 61*, 535–559.

Zhao, F. J., Zhu, Y. G., & Meharg, A. A., (2013). Methylated arsenic species in rice: Geographical variation, origin, and uptake mechanisms. *Environ. Sci. Technol., 47*, 3957–3966.

Zhao, X. Q., Mitani, N., Yamaji, N., Shen, R. F., & Ma, J. F., (2010). Involvement of silicon influx transporter OsNIP2;1 in selenite uptake in rice. *Plant Physiol., 153*, 1871–1877.

Zheng, M. Z., Li, G., Sun, G. X., Shim, H., & Cai, C., (2013). Differential toxicity and accumulation of inorganic and methylated arsenic in rice. *Plant Soil, 365*, 227–238.

CHAPTER 8

Biotechnology and Nanotechnology for Reducing the Arsenic Toxicity in Rice

TEJASVI BHATIA,[1] ANURADHA SANDHU,[1] MEGHNA,[1]
JYOTI PANDEY,[2] and HANOOR SHARMA[3]

[1]Department of Forensic Science, School of Bioengineering and Biosciences, Lovely Professional University, India

[2]Department of Chemistry, School of Physical and Decision Sciences, B. R. Ambedkar University (A Central University), India

[3]Clear Labs, San Carlos, California, United States

ABSTRACT

Rice is a popular staple food across the world. The demand and consumption of rice are increasing with elevating population. Industrialization and urbanization have led to many anthropogenic activities that pollute the environment with various toxic heavy metals. Arsenic toxicity and accumulation in rice have become a serious issue posing deteriorating effects on human, plant, and animal health. Various traditional approaches have been utilized for reducing the toxicity of rice in agriculture, but they all possess many disadvantages. The advanced technological developments have led to various biotechnological and nanotechnological approaches that greatly lower the toxicity and accumulation of arsenic in rice plants. This chapter discusses the various techniques that are used today for alleviating arsenic from rice.

Arsenic in Rice: A Global Menace. Garima Awasthi, Sudhakar Shrivastava, and Mahipal Singh Sankhla, (Eds.)

8.1 INTRODUCTION

Growing urbanization and population has resulted in various anthropogenic activities like industrialization, mining, usage of chemical fertilizers and high metal waste, etc., that have resulted in increasing heavy metal soil contamination (Ahmad et al., 2020). These trends have greatly affected the natural environment and also human health.

Ahmed et al. (2021) stated that various heavy metals such as Pb (lead), arsenic (As), zinc (Zn), cadmium (Cd), chromium (Cr), and nickel (Ni) have polluted the environment and also have affected rice production. The toxicity of these metals has resulted in lowering the production and growth of rice posing a great threat to mankind. Among them, one of the most toxic is arsenic which is responsible for causing various cardiovascular diseases, diabetes, and hypertension in humans (Wang et al., 2021). Arsenic accumulation in rice allows arsenic to enter the food chain and finally reach the human body causing various diseases. Brammer & Ravenscroft (2009) reported that a minimum of 150 million people all around the world, mainly residents of Asia, are at the verse of arsenic poisoning. It is a serious threat to crops as under arsenic stress there is a negative impact on the growth of the plant, photosynthesis of plants, and various diseases in plants like yellow coloring and fading of leaves, small root lengths, and root dysplasia are caused.

Oryza sativa L. also commonly called rice is a superfood serving carbohydrates, proteins, fats, and vitamins to the human body and therefore is an important crop upon which a large number of the population feeds (Ochoa et al., 2020). The increase of arsenic toxicity in rice is a serious concern worldwide and appropriate measures have been taken from time to time to reduce arsenic toxicity.

Biotechnology and nanotechnology play a pivotal role in reducing arsenic accumulation and providing safe rice food. Biotechnological advancements have been greatly utilized for their applications to reduce arsenic accumulation specifically in rice plants. Sun et al. (2018) reported that advanced technologies of gene modification like CRISPR/Cas-9, RNAi, gene manipulation, and editing techniques are utilized to enhance arsenic tolerance in rice plants and reduce their accumulation. Other biological methods like phytoremediation, composting, biofiltration, bioreaction, bioleaching, and biosorption are also utilized for the removal of heavy metals from the environment.

Nanotechnology has also presented a sustainable solution to reduce arsenic toxicity by producing various nanomaterials and nanoparticles. These nanoparticles possess such physiochemical properties that help in reducing the transport and uptake of these heavy metals (Liu et al., 2015). These nanoengineered materials tend to display unsaturation and high specific surface areas, which make them a promising candidate for treating water and soil pollution. Zeng et al. (2009) reported that metal oxides can be used to synthesize nanoparticles like iron oxide nanoparticles which act as an adsorbent due to their larger surface area and particle size and efficiently removed uranium. Carbon nanotubes have also been used in research for removing copper and cadmium ions from polluted water and soil (Sharma et al., 2009). Various other nanoparticles act as nano-fertilizers and nano-pesticides and are widely used for reducing arsenic accumulation and toxicity specifically in rice plants. This chapter entitles various biotechnological and nanotechnological approaches to reduce the toxicity of arsenic in rice.

8.2 BIOTECHNOLOGICAL APPROACHES FOR REDUCING ARSENIC TOXICITY IN RICE

Various biotechnological techniques have been employed to reduce arsenic toxicity and are discussed in subsections (Figure 8.1).

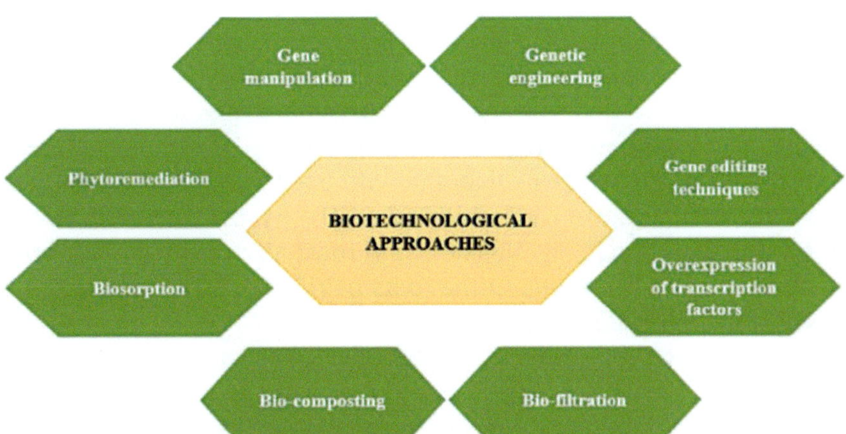

FIGURE 8.1 Various biotechnological methods are used to reduce arsenic toxicity.

8.2.1 GENETIC ENGINEERING

Plants contain arsenic in its two forms, i.e., Arsenic(III) and (V), and arsenic translocation and sequestration are highly linked with it. Thiol conjugates trap As(III) in plant roots by vacuole sequestration and As(V) comes when the conditions are aerobic. Arsenic metabolism revolves around the arsenate to arsenite reduction. Genetically engineered plants have been used for expressing the genes of arsenate reductases (ARs) and are thus used for phytoremediation. AR gene, i.e., arsC in *Arabidopsis thaliana* was overexpressed for the gene of *Escherichia coli*. This overexpression resulted in increasing arsenic hypersensitivity (Dhankher et al., 2002).

Phytoremediation can be caused due to the presence of As(III) and As(V) reductases. There is the endogenous activity of AR that also results in reducing it to arsenite which further results in slow translocation from root to shoot in arsenic (Ellis et al., 2006).

8.2.2 GENE MANIPULATION AND EDITING

8.2.2.1 AQUAPORIN GENE MANIPULATION

Various biotechnological approaches have been used for lowering arsenic accumulation in rice. One of them is the gene manipulation of aquaporin which is important for arsenic transport, root uptake, and distribution (Mukhopadhyay et al., 2014). Ma et al. (2008) reported that Lsi1 and Lis2 gene disruption led to a lowering of arsenic accumulation by negatively affecting xylem loading and root uptake. Many other researchers have also manipulated various genes to reduce the accumulation of arsenic. There was a significant lowering of As(III) in roots and straw of rice in Lsi1 mutant crops than the wild ones. In Lsi2 mutants, there was a decrease in the concentration of As(III) by 13–19% in straw and 51–63% in grain. Stele region and lateral roots also resulted in lowering the accumulation of arsenic in the roots of rice plants by knocking out OsNIP3 and OsNIP2 genes. A decrease of 16–22% in mutants and 17–22% in wild-type plants were observed when treated with OsNIP3 and OsNIP2 (Chen et al., 2017). Rice plant roots were greatly affected by gene manipulation and arsenic accumulation in roots was greatly reduced as compared to arsenic in grains.

It was also observed that the lowering of arsenic can also be achieved by increasing As(III) efflux to the rhizosphere from roots in grains of rice. Sun et al. (2018) also reported the reduction in arsenic content by the mechanism of overexpression of OsNIP1;1 and OsNIP3;3 genes in brown rice husks, grains, stems, and leaves (Figure 8.2).

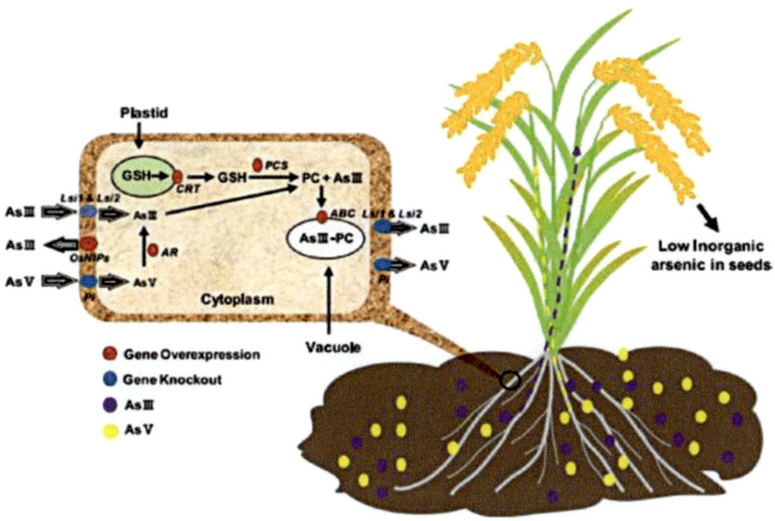

FIGURE 8.2 As(III) and As(V) uptake pathways in plant and gene manipulation for the application of reducing accumulation and toxicity of arsenic in rice.

Source: Reprinted with permission from Journal of Hazardous Materials, 408, 124751, Mawia et al. (2021); Copyright (2021), with permission from Elsevier.

8.2.2.2 PHOSPHATE TRANSPORTERS GENE MANIPULATION

Wu et al. (2011) reported that the As(V) uptake can be greatly affected by regulating and manipulating phosphate transporters in rice plants. These manipulated and mutated gene transport results in lowering the content of arsenic. Plasma membrane located OsPT1 gene functional analysis and characterization highly affected the arsenic uptake and content in rice. The low accumulation of arsenic was probably due to OsPT1's T-DNA mutants. The expression of these genes doesn't affect plant growth and development. Kamiya et al. (2013) reported that wild-type plants displayed an increase in the concentration of arsenic due to overexpression lines. OsPT8 transporter mutation also reduced AsV uptake and concentration.

Lateral roots and root tips highly express this gene which is helpful for the reduction. Ye et al. (2017) reported that the removal of the OsPT4 gene by utilizing CRISPR/Cas-9 gene-editing technology achieved a low accumulation of arsenic. OsPT4 mutant crops showed a 17–30% decrease in the wild plants in the root region. High tolerance against AsV is also one of the major properties of OsPT4 mutant plants.

8.2.2.3 TRANSCRIPTION FACTORS REGULATING ARSENIC TRANSPORTER GENES MANIPULATION

Low arsenic accumulation variety of rice can be achieved by breeding various genes that regulate the expression of arsenite and arsenate positively or negatively. Various transcription factors regulate the transportation of arsenic in plants that can be genetically manipulated. The lower and higher affinity of phosphate transporter genes OsPT2 in the plasma membrane is highly regulated by transcription factor OsPHF1. By removal of these OsPHF1 transcription factors can help in achieving arsenic resistance by increasing OsPT2 and OsPT8 retention in the endoplasmic reticulum (Chen et al., 2011). Wang et al. (2017) reported that the transcription factor OsARM1 (arsenite-responsive MYB1) regulated Lsi1, Lsi2, and Lsi6 gene expression in rice crops. Increased resistance to As(III) was achieved by the removal of this gene by enhancing the AsIII transporter gene upregulation. Overexpression lines also displayed downregulation of arsenic transporter genes and high arsenic sensitivity in mutant plants. The lines of overexpression showed lower content of arsenic in mutated plants than wild and OsARM1 mutants showed high arsenic in shoots. Seven days study was conducted and it displayed the overexpression of OsARM1 and its effect on arsenic content in the root and shoot of rice. Phosphate transporters and AsV content in rice are also regulated by another transcription factor named OsWRKY28 (Wang et al., 2018a). Reproductive organs, root tips, and lateral roots of plants express this gene highly and its removal caused a significant lowering in the content of arsenic. Around 16–31% lowering of arsenic content was observed in the mutant type and 28–34% lower in Pi transporters. It was concluded that low and high levels of gene expression can be obtained by regulating the transcription factor (Table 8.1).

TABLE 8.1 Genetic Manipulation of Various Genes and Their Results

Name of the Gene	Category	Genetic Manipulation	Outcome	References
Lsi1 (OsNIP2;1)	Aquaporin transporters	Gene knockout	Reduced accumulation of arsenic in rice.	Ma et al. (2008)
AtNIP1;1	Aquaporin transporters	Gene knockout	Tolerance against arsenic increased and toxicity decreased.	Kamiya et al. (2009)
OsPht1;8	Phosphate transporters	Gene knockout	Tolerance against AsV increased.	Shin et al. (2004)
OsHAC1;1 and OsHAC1;2	Arsenate reductase	Gene overexpression	Lower arsenic accumulation in grain of rice and its high efflux to the outer environment.	Shi et al. (2016)
OsABCC1	ABC transporter	Gene overexpression	Helps in obtaining higher arsenic tolerance.	Song et al. (2014)

8.2.3 INCREASING VACUOLAR SEQUESTRATION AND ARSENIC DETOXIFICATION

As(III) is formed by the reduction of As(V) by the action of AR inside plant cells that results in the formation of AsIII-PC and AsIII-GSH complexes. Lowering arsenic content has been reported by manipulating the Phyto chelation modulated genes and arsenic reductase in rice crops. Enriched arsenate tolerance and lowered accumulation were achieved by OsHAC1;1 and OsHAC1;2 genes overexpression. Paddy soils polluted with arsenate were taken for an experiment by Shi et al. (2016) where overexpressing lines resulted in a lowering of 20% arsenic content than other wild-type plants owing to increased efflux of As(III).

Xu et al. (2017) reported that the detoxification of As(V) and its accumulation in crops of rice were also affected by OsHAC4 gene editing. OsHAC4 is present in the cytoplasm and nucleus of the plant and its expression is mainly in the exodermis and epidermis of the root. The tolerance of arsenate was also found to be enhanced. Shoot and roots were found to be having lowered arsenic content by 39.2–42.1% in roots and shoots by 38.2–40% due to overexpression of this gene. OsHAC4 activity function loss led to increased arsenic content in shoots, reduction of efflux of As(III) into the external environment, and reduced arsenate reduction in roots of rice.

As(III) tolerance was enhanced and accumulation was lowered by utilizing the OsGrx_C7 gene and manipulating it for its overexpression (Verma et al., 2020).

8.2.4 RICE BREEDING

Breeding of rice crops has also been done in literature to reduce the concentration of arsenic in plants. Syu et al. (2015) reported that specific cultivators are selected that restrict the biological accumulation of arsenic in grains of rice, thus leading to at least 20 to 30-fold reduced arsenic content than the others. The accumulation, translocation, or uptake can be reduced or blocked by using these cultivators as they are special low arsenic-containing breeds of rice (Norton et al., 2012). So, by genetic engineering technology, special breeds of rice can be developed that are tolerant towards arsenic.

8.2.5 BIOSORPTION

Biosorption is a part of explicit biomass that doesn't attempt to draw in and tie substantial metal particles to aqueous solutions (various levels of cleaning) yet Biomass assumes the part of synthetic compounds, for example, particle trade in natural qualities. This condition has been demonstrated to rely upon the construction of the cell mass of biomass. Its contrary interaction is called desorption which performs "return of promotions and reactivation of adsorbent." The process of biosorption is classified into two phases: solid and liquid. The solid phase is sorbent or biological substance and the liquid phase is solvent, usually water, solvent is removed in different ways as they are drawn towards sorbent. This process continues until a condition of "balance is reached between the size of the strongly fortified sorbate and its liquefied piece of the arrangement. This cycle goes as follows: first, the connection between the metal and the dynamic gatherings reflected in the surface cells of the sorbent prompts the publicizing of metal particles over the outer layer of the phones. The last option offers an assortment of possible restricting particles for metal particles." The conveyance between the two stages depends on the firm force of the sorbent. Contrasts in cell divider arrangement in various gatherings of microorganisms (green growth, parasites, microbes, and so forth) influence the sort and number of metal particles. This first stage, called inactive biosorption, is characterized as "the unique equilibrium of cement adsorption-desorption: Metal particles bound to the phone can be eliminated by different particles, chelating specialists, or acids." It happens because of at least one of the accompanying metal-restricting cycles: particle trade, lifeless precipitation, ligand trade, holding, and intricacy. Second, the dynamic piece of biosorption is started as iron particles arrive at the cells through their films. Biosorption can be accomplished in two ways: living cells or non-living cells. In any case, living cells can commit suicide with publicized cells that lead to cell demise. What's more, living cells require the acquisition of supplements that will expand the BOD and COD of wastewater. Accordingly, lifeless cells stay an entirely reasonable choice as they require negligible change and can be reused.

8.2.6 BIOREACTOR

"Bio-reactor" is a typical name for a framework that debases groundwater and soil-containing microorganisms. The reactors can be open frameworks, like an inherent wetland (characterized as a different innovation), or an encased room. This depiction just portrays the most recent. In contrast to normal relief and *in-situ* bioremediation, bioreactors can forestall and control normal intricacies of native microorganisms that are idle as well as low degrees of conventional microbial. There are a few sorts of bioreactors. The most well-known bioreactor is utilized to treat wastewater. Debased groundwater is conveyed to the air admission where microorganisms pollute creatures, shaping expendable or reused mud. The subsequent kind uses a turning natural framework. Microorganism figures fill in a lattice, which is turned into a reactor. One strategy is utilizing a full bed. The tank is loaded with medium help, which gives a huge region to microbial development. One framework utilizes soil slurry bioreactor innovation to obliterate soil containing dynamite (TNT) and Royal Demolition Explosive (RDX). Bioreactors can likewise be introduced *in situ* (i.e., locally). Determined bioreactors are called bio-plugs. A divider, like a reactor block divider that can be introduced, can be introduced. Level bioreactors are called bio-conductors. These techniques advance debasement as sullied groundwater goes through the reactor. This innovation has been utilized effectively to fix natural mixtures in broken underground stockpiling tanks and modern regions.

Bioreactors are essentially used to treat unstable natural mixtures (VOCs: volatile organic compounds) and fuel hydrocarbons in soil and groundwater. This cycle doesn't function admirably for pesticides. In one application, the idea was utilized to treat soils containing TNT and RDX. In the research center, it worked under high-impact and anaerobic conditions, and there was a critical decrease in fixation. Middle-of-the-road items have additionally been diminished. *In-situ* bioreactors can likewise be utilized to give co-metabolite to the deterioration of hurtful items created during the decay interaction in a portion of the chlorine-containing solvents. This sort of bioreactor contains hereditarily changed life forms that make minerals for living organic entities. Microscopic organisms are caught in a natural help climate. The *in-situ* immobilized bioreactor framework can be utilized related to a dissipation framework.

8.2.7 PHYTOREMEDIATION

Plants have been utilized for the phytoremediation of poisonous metals and metalloids; however, this cycle has been slow and to a great extent ineffectual because of the phytotoxicity of weighty metals in plants (Dhankher et al., 2012). Normal hyperaccumulators for weighty metals are likewise accessible. In any case, it doesn't have the fundamental biomass required for phytoremediation to work and be restrained in certain geo-climatic conditions. Hereditary designing strategies to further develop plant resistance and an assortment of poisonous metals have incredible potential for phytoremediation. Also, numerous new investigations utilizing omics innovation including genomics, transcriptomics, proteomics, and metabolomics have been performed to explain the hereditary choices and instruments engaged with the resistance of substantial metals and metalloids to plants. Biotechnological strategies are as of now being utilized to consolidate substantial metals and metalloids like mercury (Hg), cadmium (Cd), lead (Pb), selenium (Se), copper (Cu), and arsenic (As). Three fundamental biotechnological techniques utilized in plant designing to work on the phytoremediation of substantial metals and metalloids are as follows:

- The administration of metal/metalloid qualities and catch frameworks;
- To advance the creation of iron and metalloid ligand;
- The transformation of metals and metalloids into less poisonous and adaptable structures (Kotrba et al., 2009).

8.2.8 ACTIVATED SLUDGE

A multi-chamber reactor uses highly concentrated microorganisms for the degradation of organisms and the removal of nutrients from water for the production of high-quality pollutants. To keep the sediments suspended and maintain aerobic conditions, an ideal oxygen supply is required. It contains bacterial flocs which are blended and suspended with tainted water in circulated air through the tank. Microscopic organisms use poisons to develop and change over into energy, CO_2, new cell substances, and water. Enacted mud frameworks are a kind of fixed development and are utilized in customary wastewater treatment plants to treat practically all effects of sullied water as long as it doesn't spoil. A pre-exercise-based recuperation unit, a post-settlement unit (clarifier) in which the dynamic mud is re-scattered in circulated air through the tank, and the treatment of abundance mud, are required for legitimate treatment. This cycle is exceptionally mechanical and along these lines adjusted fundamentally to unified frameworks where power, spare pieces of hardware, and talented staff are accessible. However, long as the reactor is working appropriately, the best evacuation of natural and strong particles can be accomplished, albeit the expulsion of the microorganism is low.

8.2.9 BIO-COMPOSTING

Small aerobic microorganisms convert the other organisms and organic matter into compost calling the phenomenon composting. It has a thermophilic, mesophilic, and cooling phase. Around 40–50% reduction in solid waste can be achieved by this method. The thermophilic phase produced metabolic heat that causes bacteria to break down in return reducing the harmful contaminants. Soil properties are also improved by utilizing this method as it acts as a fertilizer. Heavy metals deteriorate the health of the soil, human, plant, and the whole environment. These heavy metals like arsenic are increased by various processes such as biomagnification. Clinoptilolites and natural zeolites act as a supplement to soil and are efficient for removing heavy metals. The chemical properties of heavy metals are altered by the addition of clinoptilolite stone in sewage which in turn lowers the bioavailability and mobility of arsenic and other heavy metals.

8.2.10 BIO-FILTERING

Microorganisms are utilized in the bio-filtration interaction to scatter foreign substances in the polluted stream. Microorganisms fill in the bio-film surface of the medium or are suspended in the water stage around the middle particles. The bed channel contains non-practical materials that guarantee enormous spaces of connection just as the arrangement of extra supplements. The general proficiency of the natural channel is to a great extent dictated by the properties and attributes of the supporting material, including toughness, thickness, water maintenance limit, and the capacity to deal with a few microorganisms. Key execution boundaries of the biofilter channel incorporate microbial infusion, medium pH, temperature, moderate moistness, and supplement content. As the sifting system advances, microorganisms (anaerobic, protozoan, facultative, bacterial, protozoa, aerobic, and fungal) are progressively evolved on the channel surface and structure an organic film or flimsy layer known as a biofilm. A significant point in the adequacy of the biofilter is to control and keep up with sound biomass on the outer layer of the channel. As the capacity of a natural channel is exceptionally reliant upon microbial action, a nonstop wellspring of substrates (natural matter and supplements) is expected to work reliably and viably, albeit some chemoautotrophic microorganisms might utilize lifeless synthetics as a wellspring of energy. The productivity of removing bio-channels is constrained by different boundaries, for example, pH, temperature, O_2 content, beginning immersion of poisonous toxins, and so forth Powerful evacuation might be upgraded by substance adjustment of a media channel or hereditary change of microorganisms (Srivastava et al., 2008).

8.3 NANOTECHNOLOGY APPROACHES FOR REDUCING ARSENIC TOXICITY IN RICE

Nanotechnology is the manipulation, utilization, and processing of the nanoscale dimension materials commonly known as nanomaterials and nanoparticles having tuneable and desirable properties (Roduner, 2006). It is an emerging field that exhibits various applications in agriculture. It displays sustainable agriculture approaches that benefit the environment also. Younas et al. (2020) stated that nanotechnology is an advanced technology that has enhanced rice agriculture and production and due to this, it has gained a lot of attention worldwide. Nanomaterials like nano-fertilizers, nano-adsorbents, and nano-pesticides are synthesized time and again for the nano-remediation of heavy metals like arsenic (Maity et al., 2020). Khan & Rizvi (2017) explained that utilizing nanotechnology-based nanomaterials results in a controlled and gradual release of nutrients into the soil which enhances the soil conditions.

Arsenic toxicity is a worldwide problem and there are various approaches in nanotechnology to deal with it starting from the utilization of metal oxides to reduce arsenic toxicity. Nowadays, nano-metal oxides are utilized as they exhibit more desirable properties like zinc oxide, iron (hydro)oxides, titanium dioxide, copper oxide, and various other metal oxide nanoparticles. These act as an adsorbent and remove arsenic and other heavy metals from the environment (Lata & Samadder, 2016).

The adsorption of these nanoparticles occurs when they accumulate in soil particles and other microorganisms present in the soil. Soils pH condition, organic matter, minerals in the soil, texture of the soil, surfactants, and ionic strength are the conditions altering the adsorption (Joo & Zhao, 2017; Gao et al., 2019; Qian et al., 2020) (Figure 8.3).

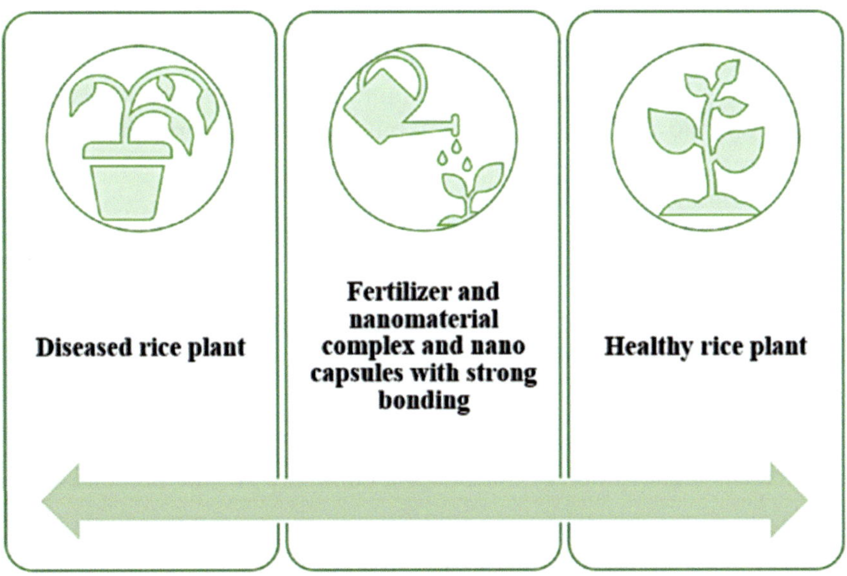

FIGURE 8.3 Role of nanotechnology to enhance rice production.

8.3.1 NANO-PHOTO-CATALYSIS

Nano-photocatalysis has been widely used to treat waste in the field of environment and environmental safety, as it has many advantages, such as those low cost, amazing stability, high photocatalytic activity, etc. Various methods, including ion exchange microorganisms and adsorption, have been used to treat feces. However, these methods are limited due to their sophisticated technology, high cost, risk of secondary contamination, and poor degradation performance. Compared with other biomedical waste management methods. Nano-photocatalysis is considered one of the most innovative approaches to energy use, and environmental issues. The advantages of a photocatalysis strategy are as follows:

i. Photocatalysis offers a respectable alternative to traditional energy-based therapies (e.g., ultrafiltration and reversible osmosis) with the ability to use non-polluted and renewable solar energy.
ii. Promotes the production of innocent products, in contrast to conventional therapies where contamination only transfers from one stage to another.
iii. The process may be used to eliminate a variety of hazardous and hazardous compounds in different streams of wastewater.
iv. Requires the application of minimally invasive chemicals and can operate under conditions of mild reaction during the moderate reaction.
v. Minimum secondary waste generation.
vi. Can also be used in solid phase (soil), gas phase (hydrogen generation), and water treatment.

The reliance on photocatalytic activity in the following conditions affected their application:

- cost-sharing;
- the face-to-face transfer charge needs to be improved;
- the reorganization of the network company may be prevented.

8.3.2 NANO-ENGINEERED NANOPARTICLE

Nanoparticles designed for shape, size, and surface structure offer exceptional performance that includes attractive behavior, enhanced power, enhanced thermal conductivity and electricity, and the controlled release of host molecules. These developments have opened up applications for biomedicine, energetic nanomaterials, and active nanocomposites. Various nanoparticles are synthesized for various purposes. Developed NPs are being studied extensively for use in agricultural and environmental sustainability. 'Phyto nanotechnology' collectively combines plant production and science with nanotechnology for utilization in various applications. The use of NPs in plant systems to improve fertilizer, pesticides, and nerves is aimed at growing these substances to meet plant needs and reduce waste. Some of them are discussed below in specificity with the arsenic reduction in rice plants.

8.3.2.1 COPPER OXIDE (CUO) NANOPARTICLES

Plant growth and development highly depend on micronutrients like copper. It helps in maintaining the optimum levels required for photosynthesis in plants and various other processes (Yruela et al., 2013). Various reports support the utilization of copper oxide nanoparticles for reducing arsenic toxicity in rice plants and their use as nano-fertilizers. Liu et al. (2018) investigated the combined effects of copper oxide nanoparticles and arsenic on the germination and growth of seedlings in rice plants (*Oryza sativa japonica*). The 18 days experiments were conducted to see the effects of copper nanoparticles of various concentrations (0, 0.1, 1.0, 10, 50, and 100 mg/L) and arsenic. Arsenic treatment caused seed germination inhibition and various negative effects on plant growth and as a result, the rice productivity was adversely decreased during high As concentrations. The nanoparticles of copper oxide were then utilized for the treatment and a negative correlation was observed between copper oxide nanoparticles and arsenic content. The data in the experiment supported that the arsenic uptake can be reduced by using copper oxide nanoparticles.

The utilization of CuO nanoparticles to observe the arsenic effects on the life cycle and growth of rice plants has been evaluated by Liu et al. (2018). Dehusked grains were taken and treatment with CuO nanoparticles with arsenic showed a reduction of arsenic accumulation by 35% as compared to other arsenic treatments. The world health organization has concluded the safe value of arsenic in white rice safe for human consumption to be 200 ng/g (Åkesson et al., 2012) and after treatment with copper oxide nanoparticles, the arsenic accumulation was observed at 128 ± 15.4 ng/g in dehusked grains. Martinson & Reddy (2009) also reported that the main factor responsible for the reduction of arsenic accumulation is credited to the adsorption phenomenon of AsV that occurs on the surface of copper oxide nanoparticles that in turn results in reduced bioavailability and mobility of arsenic in the soil. Wang et al. (2019) also reported the influence of copper oxide nanoparticles on arsenic accumulation in rice plants. Reduction in both root and shoot total arsenic content occurred by 23% and 45%, respectively, due to the activity of copper oxide nanoparticles which displayed the on-surface adsorption of two iAs species contributing to reducing the arsenic accumulation in tissues of rice. Synergistic phytotoxicity of copper nanoparticles and arsenic also resulted in reducing the uptake of arsenic. Wang et al. (2013, 2015) reported that the Copper oxide nanoparticles and AsIII co-existence results in lowering 71% and 45% of the AsIII and AsV species, respectively in roots of rice when compared with those treated with only AsIII (Table 8.2). The main factors for more reduction of AsIII than AsV are as follows:

- Redox reactions between the AsIII and Cu^{2+} ions resulted in releasing AsV and Cu^{+} ions.
- ROS induction due to copper oxide nanoparticles resulted in the oxidation of AsIII to AsV.

TABLE 8.2 Effect on Arsenic Accumulation and Toxicity by Application of ZnO and CuO Nanoparticles

Nanoparticles	Media for Growth	Used Concentration	Results	References
Copper oxide (CuO)	Soil	0.1–100 mg/L	35% reduced accumulation of arsenic was observed.	Liu et al. (2018)
Copper oxide (CuO)	Hydroponic	100 mg/L	Roots displayed a 23% decrease and shoots displayed a 45% decrease in the content of arsenic.	Wang et al. (2019)
Zinc oxide (ZnO)	Hydroponic	100 mg/L	Zinc oxide nanoparticles helped in a total reduction of 68% and 72% of arsenic content in rice roots.	Wang et al. (2018b)
Zinc oxide (ZnO)	Hydroponic	100 mg/L	Here zinc oxide nanoparticles acting as nano-fertilizers were able to reduce the accumulation of arsenic in rice roots and shoots.	Ma et al. (2020)

8.3.2.2 ZINC OXIDE (ZNO) NANOPARTICLES

As accumulation decrease was observed in an experiment where a seedling of rice was grown using hydroponics and ZnO nanoparticles with AsIII, and ZnO nanoparticles with AsV combinations were used for treatment as compared to that when exposed to alone AsIII or AsV (Wang et al., 2018). ZnO NPs + AsIII and ZnO NPs + AsV combinations resulted in a total reduction of 72% and 68% respectively after exposure to total arsenic content which was way higher than treatment with AsIII and AsV same concentration alone. Two iAs species present in the roots of the plant decreased in their concentrations when treated with ZnO nanoparticles. AsIII showed a larger decrease than that AsV. Many factors influence this reduction of arsenic in roots such as:

- When ZnO nanoparticles and iAs species co-occurred, they exhibited a negative phyto-effect on the development of the root of the plant that resulted in inadequate operation of transporters of arsenic.
- Reactive oxygen species generation occurred due to the presence of zinc oxide nanoparticles which resulted in electron depletion required for arsenic reduction by oxidizing all exudates of the root.
- AsV reduced reduction inside root cells of plants.
- Reduction in arsenic mobility due to the presence of adsorbent on the surface of zinc oxide nanoparticles.
- And by also reducing the nutrient solution bioavailability.

ZnO NPs and AsIII treated rice shoots exhibited more arsenic content as compared to those treated with ZnO NPs and AsV since both have different redox statuses and use distinct uptake and metabolic pathways. There was no effect on zinc uptake by the interaction of nano-engineered nanoparticles and arsenic species (Wang et al., 2018).

Wu et al. (2020a, b) reported there was a total reduction of 72.3% and 56.6% in arsenic concentrations in roots and shoots of rice, respectively using ZnO nanoparticles. It was observed that these zinc oxide nanoparticles highly affected the early growth, germination, arsenic uptake, and various other biochemical processes in rice plants. The decreased arsenic mobility and its bioavailability are the main causes that result from the adsorption of arsenic on zinc oxide nanoparticles that result in a low total accumulation of arsenic in rice.

Another researcher experimented to observe the reduction of arsenic and cadmium in rice utilizing ZnO nanoparticles. The seeds were exposed with and without zinc oxide nanoparticles for observation. The results displayed that the utilization of ZnO nanoparticles led to a decrease in arsenic accumulation by 39.5% in the roots of rice plants and 60.2% in shoots as compared to those seeds treated without it (Ma et al., 2020). These reports suggested that zinc oxide nanoparticles display themselves as an efficient agrichemical or a nano-fertilizer that helps in the reduction and accumulation of arsenic in rice (Figures 8.4(a) and (b)).

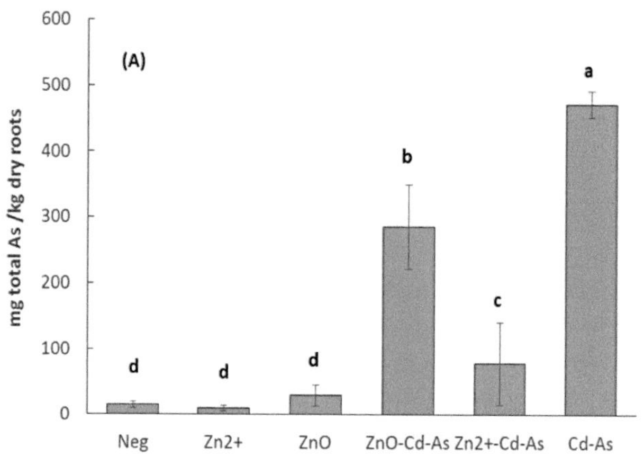

FIGURE 8.4(A) Different treatments display total arsenic concentration in rice roots.

Source: Reprinted from Chemical Engineering Journal, 384, 123802, Ma, Sharifan, Dou, & Sun (2020); Copyright (2020), with permission from Elsevier.

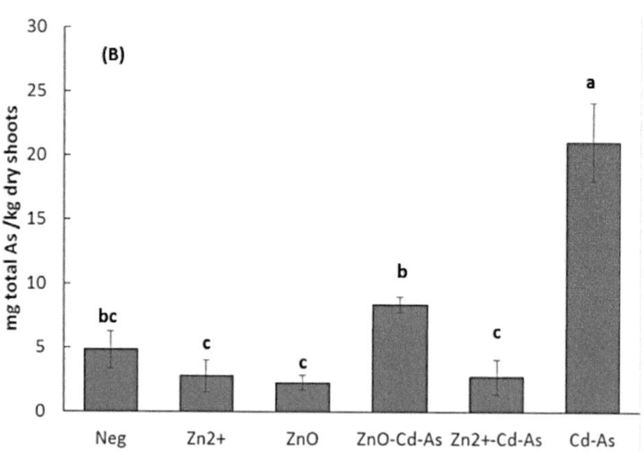

FIGURE 8.4(B) Different treatments show total arsenic concentration in rice shoots.

Source: Reprinted from Chemical Engineering Journal, 384, 123802, Ma, Sharifan, Dou, & Sun (2020); Copyright (2020), with permission from Elsevier.

8.3.2.3 MANGANESE DIOXIDE (MNO2) NANOPARTICLES

Accumulation and transport of arsenic with the treatment of nano-fertilizer manganese dioxide nanoparticles (MnO_2) were studied in rice plants. The experiment displayed that there was a negative correlation in the distribution of arsenic and manganese in various organs of the plant. A reduction of 65.4% in arsenic content was seen in rice grains when treated with MnO_2 nanoparticles than those of arsenic-treated plants only. The improved oxidation of AsIII to AsV is the main reason for the reduction of As in rice grains when treated with manganese dioxide nanoparticles.

Researchers studied arsenic accumulation in paddy soils that were contaminated with it with the help of nanostructured α-MnO_2. Critical negative relationships were seen among arsenic and manganese in rice as there was an increase of Mn in soil and a decrease in arsenic (Li et al., 2019). The build-up of arsenic was significantly low in nanoengineered treated roots and husks of rice plants in comparison with those treated with arsenic only. *In-situ* oxidation to AsV from AsIII happened when the plants were supplemented with nanorods of α-MnO_2 which increased the rate of adsorption of AsV.

Ehlert et al. (2014) also studied the role of manganese dioxide nanoparticles in arsenic accumulation and stated that they help in the reduction of arsenic as iAs promote Fe^{3+} oxidation from Fe^{2+} which created iron hydroxide which acts as sorbent and reduces the bioavailability and mobility of arsenic in rice.

8.3.2.4 SILICON DIOXIDE (SIO2) NANOPARTICLES

Silicon dioxide also acts as a nano-fertilizer and is helpful in the reduction of arsenic content from rice plants. Straw and grains of rice were reported to show low arsenic toxicity when treated with silicon fertilizers (Li et al., 2009). Detmann et al. (2012) reported that silicon is a very important nutrient that helps in various functions of plant-like remobilization of amino acids, increase in spikelet per panicle in the plant, and increase in the percentage of spikelet.

Pot and field experiments showed a decrease of 28% and 40% in arsenic concentration using silicon dioxide nanoparticles in the grain of rice. The concentration of arsenic in roots decreased by 12.2% and 22.3% and in shoots by 53.2% and 41.3% in these experiments. Silicon uptake by plants follows the same pattern as arsenic uptake so there was competition among both due to reduced lipid peroxidation in the roots of rice. The following factors resulted in reducing arsenic accumulation (Ma & Yamaji, 2006):

- Xylem unloading became the main factor that reduced arsenic accumulation.
- Deposition of silicon on the cell wall resulted in reducing the translocation of arsenic, resulting in reducing its accumulation in rice.
- Inhibition of Lis1 and Lis2 gene expression in roots of rice due to the application of nanoengineered silica.
- The increase in pectin content and thickness of the cell wall also contributed to reducing arsenic accumulation (Cui et al., 2020).

Other researchers also reported a decrease in the accumulation and phytotoxicity of arsenic in rice by utilizing nano-fertilizers of silica and silicon nanoparticles (Liu et al., 2014; Mawia et al., 2021). Grains of rice were taken for the experiment to be tested for arsenic toxicity by utilizing silicon dioxide nanoparticles. The foliar application of silicon dioxide nanoparticles increased the yield and significantly lowered the arsenic toxicity (Figure 8.5).

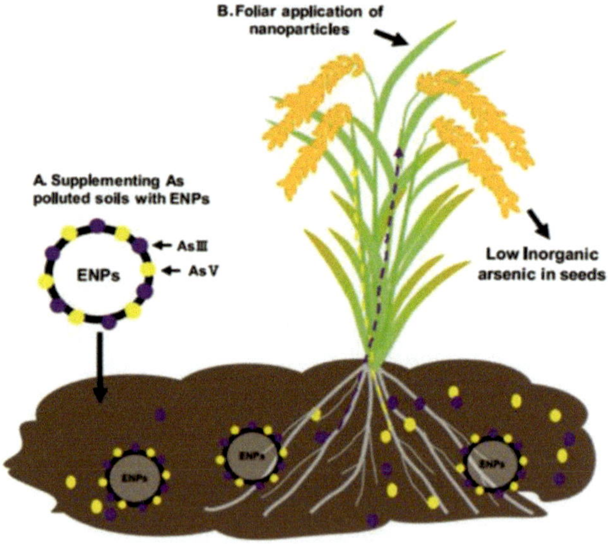

FIGURE 8.5 Utilization of silicon dioxide nanoparticles in alleviating arsenic accumulation in rice plants.

Source: Reprinted with permission from Journal of Hazardous Materials, 408, 124751, Mawia et al. (2021); Copyright (2021), with permission from Elsevier.

8.3.2.5 IRON OXIDE NANOPARTICLES

Removal of arsenic from soil and water contaminated with arsenic has been done utilizing iron nanomaterials. Nano-zerovalent iron and Nano-Fe_3O_4 have been used by Huang et al. (2018) to study the accumulation of arsenic in rice. At both high and low concentrations of arsenic, the utilization of nanoengineered nanoparticles displayed an effective lowering of arsenic accumulation in rice. Adsorption occurred in growth solution due to nanoparticles presence thus lowering the bioavailability of arsenic in rice plants. All the results displayed that adsorption on the surface of nanoparticles is the main factor that reduces the accumulation and uptake of arsenic under various conditions (Table 8.3).

TABLE 8.3 Effect on Arsenic Accumulation and Toxicity by Application of Various Nanoparticles Acting as Nano-Fertilizers

Nanoparticles	Media for Growth	Used Concentration	Results	References
Manganese dioxide (MnO_2)	Soil	0.2–2.0%	A 65% reduction in the content of arsenic in grains of rice was observed.	Zhou et al. (2015)
Alpha-manganese dioxide (MnO_2)	Soil	0.2–2.0 W/W	Significantly less arsenic content was found in the husk and roots of brown rice.	Li et al. (2019)
			Total As.	
Silicon dioxide (SiO_2)	Hydroponic and soil	5 mM	Decreased total As in the grains by 28% and 40%	Liu et al. (2014)
Silicon dioxide (SiO_2)	Hydroponic	0.1–10 mM	Expression levels of Lsi1 and Lsi2 were reduced by using these nanoparticles thus leading to reduced arsenic accumulation.	Cui et al. (2020)
Iron oxide (Fe_3O_4 and zero-valent iron)	Hydroponic	0–4 mg/L	Roots and shoots of rice showed very little arsenic toxicity and high tolerance.	Huang et al. (2018)
Cerium dioxide (CeO_2)	Hydroponic	100 mg/L	Rice grains exhibited no significant changes in terms of arsenic content.	Wang et al. (2018b)
Titanium dioxide (TiO_2)	Hydroponic	0–100 mg/L	40–90% lowered bioaccumulation of arsenic in seedlings of rice.	Wu et al. (2021)
Montmorillonite clay containing nanoparticles in rice husk biochar.	Soil	0–25% W/W	73% decrease in arsenic concentration.	–

8.3.2.6 CERIUM OXIDE (CEO2) NANOPARTICLES

Various other nanoparticles like cerium oxide nanoparticles have also been utilized in literature to reduce arsenic accumulation and toxicity. Rossi et al. (2018) reported that there was a high concentration of AsV in shoots of rice but a low concentration of AsIII due to the co-existence of cerium nanoparticles and arsenic. This copresence also presented a high increase of cerium concentration in rice shoots as the transportation of cerium ions is efficient when present in nanoparticle form to shoots.

The utilization of AsIII in treatment resulted in lowering the AsV concentration in the roots of rice plants due to the co-existence of cerium oxide nanoparticles with AsV leading to root exudates excretion triggering AsV reduction into AsIII (Wang et al., 2018b).

8.3.2.7 TITANIUM DIOXIDE (TIO_2) NANOPARTICLES

Alleviation in the accumulation of arsenic in rice can be reduced by utilizing nanoparticles of titanium dioxide (Wu et al., 2021). Hydroponic experiments were performed to study the effect of TiO_2 nanoparticles on rice plants for a duration of seven days where different concentrations of nanoparticles were used. The nanoparticles of titanium were able to reduce the bioaccumulation of As without affecting the growth of the rice plant. The robust sorption phenomenon resulted in reducing the As content by 40–90% in rice. In the absence of these nanoparticles, a strong barrier in the form of iron plaque was found that reduced the uptake of arsenic. But the utilization of titanium dioxide nanoparticles greatly decreased the iron plaque also and reduced the accumulation of arsenic strongly making them more efficient.

Wu et al. (2020) also investigated different titanium nanoparticles like BT-bulk titanium dioxide nanoparticles, NAT-anatase titanium dioxide nanoparticles, and NRT-rutile titanium dioxide nanoparticles to see their effect on arsenic accumulation. It was found that NRT reduced arsenic accumulation by 88.4% and NAT by 39.7% in rice which was higher than control samples. Arsenic content in the shoot decreased with the help of NRT 90% and NAT by 14%. Iron plaque retention and adsorption of titanium dioxide nanoparticles have proven themselves to be the principal factors to reduce the accumulation of arsenic (Figure 8.6).

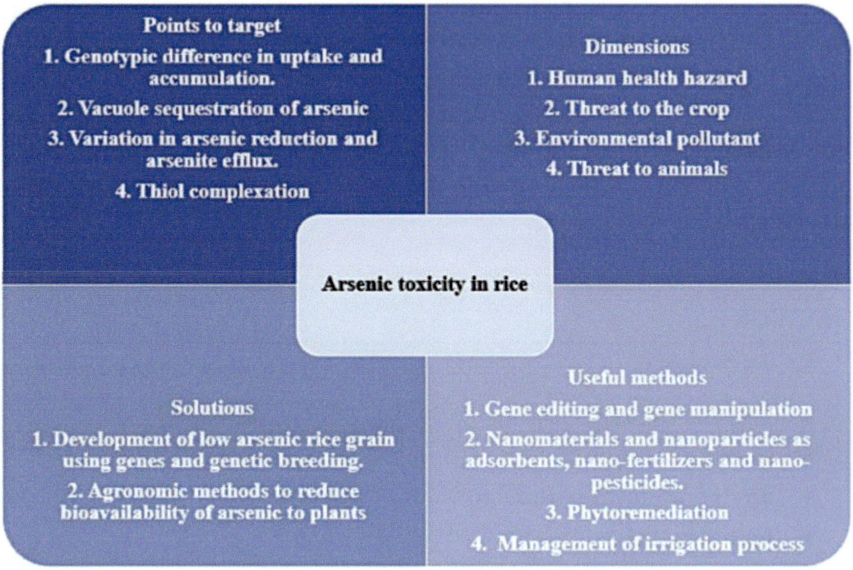

FIGURE 8.6 Overview of arsenic toxicity in rice plants and its solutions.

8.4 SUMMARY

The demand for rice is ever-increasing with the increasing population, but various biotic and abiotic factors pose a threat to rice production. Various anthropogenic and industrial activities have led to environmental pollution and toxicity. One of the most toxic outcomes is arsenic accumulation in rice which serves as a threat to humans, animals, and plants causing different diseases in each one. Its removal has been done by various conventional technologies used in agriculture, but they all resulted in partial removal, high expense, and toxic sludge generation. Nowadays with evolving technology various biotechnological and nanotechnological approaches have been coming to light for reducing the accumulation and uptake of arsenic in rice crops. Genetic manipulation, genetic editing, and various other gene knockout, overexpression strategies of biotechnology have shown great influence on arsenic tolerance in rice. Nanotechnology has also given various nanomaterials and nanoparticles such as zinc oxide, cerium oxide, iron oxide, titanium dioxide, and silicon dioxide that acts as nano-fertilizers and nano-pesticides. These nano-engineered particles owing to their particle size and high surface area act as good adsorbents and consequently decrease the accumulation and uptake of arsenic in rice. These various biotechnology and nanotechnology advancements have presented a feasible, economical, and sustainable solution to this growing problem of arsenic toxicity and in the near future, many more techniques are expected.

KEYWORDS

- **arsenic toxicity**
- **biotechnology**
- **environmental pollution**
- **heavy metals**
- **nano-engineered particles**
- **nano-pesticides**
- **nanotechnology**
- **rice**

REFERENCES

Ahmad, P., Alam, P., Balawi, T. H., Altalayan, F. H., Ahanger, M. A., & Ashraf, M., (2020). Sodium nitroprusside (SNP) improves tolerance to arsenic (As) toxicity in *Vicia faba* through the modifications of biochemical attributes, antioxidants, ascorbate-glutathione cycle and glyoxalase cycle. *Chemosphere, 244*, 125480.

Ahmed, T., Noman, M., Manzoor, N., Shahid, M., Abdullah, M., Ali, L., & Li, B., (2021). Nanoparticle-based amelioration of drought stress and cadmium toxicity in rice via triggering the stress responsive genetic mechanisms and nutrient acquisition. *Ecotoxicology and Environmental Safety, 209*, 111829.

Åkesson, M. T., Point, C. C., & Di Caracalla, V. D. T., (2012). *Proposed Draft Maximum Levels for Arsenic in Rice (AT STEP 3).* Codex Alimentarius Commission. Joint FAO/WHO Food Standards Program, Codex Committee on Contaminants in Foods: Maastricht, *Netherlands*.

Brammer, H., & Ravenscroft, P., (2009). Arsenic in groundwater: A threat to sustainable agriculture in South and South-east Asia. *Environment International, 35*(3), 647–654.

Chen, J., Liu, Y., Ni, J., Wang, Y., Bai, Y., Shi, J., & Wu, P., (2011). OsPHF1 regulates the plasma membrane localization of low-and high-affinity inorganic phosphate transporters and determines inorganic phosphate uptake and translocation in rice. *Plant Physiology, 157*(1), 269–278.

Chen, Y., Sun, S. K., Tang, Z., Liu, G., Moore, K. L., Maathuis, F. J., & Zhao, F. J., (2017). The Nodulin 26-like intrinsic membrane protein OsNIP3; 2 is involved in arsenite uptake by lateral roots in rice. *Journal of Experimental Botany, 68*(11), 3007–3016.

Cui, J., Li, Y., Jin, Q., & Li, F., (2020). Silica nanoparticles inhibit arsenic uptake into rice suspension cells via improving pectin synthesis and the mechanical force of the cell wall. *Environmental Science: Nano, 7*(1), 162–171.

Detmann, K. C., Araújo, W. L., Martins, S. C., Sanglard, L. M., Reis, J. V., Detmann, E., & DaMatta, F. M., (2012). Silicon nutrition increases grain yield, which, in turn, exerts a feed-forward stimulation of photosynthetic rates via enhanced mesophyll conductance and alters primary metabolism in rice. *New Phytologist, 196*(3), 752–762.

Dhankher, O. P., Li, Y., Rosen, B. P., Shi, J., Salt, D., Senecoff, J. F., & Meagher, R. B., (2002). Engineering tolerance and hyperaccumulation of arsenic in plants by combining arsenate reductase and γ-glutamylcysteine synthetase expression. *Nature Biotechnology, 20*(11), 1140–1145.

Dhankher, O. P., Pilon-Smits, E. A., Meagher, R. B., & Doty, S., (2012). Biotechnological approaches for phytoremediation. In: *Plant Biotechnology and Agriculture* (pp. 309–328). Academic Press.

Ehlert, K., Mikutta, C., & Kretzschmar, R., (2014). Impact of birnessite on arsenic and iron speciation during microbial reduction of arsenic-bearing ferrihydrite. *Environmental Science & Technology, 48*(19), 11320–11329.

Ellis, D. R., Gumaelius, L., Indriolo, E., Pickering, I. J., Banks, J. A., & Salt, D. E., (2006). A novel arsenate reductase from the arsenic hyperaccumulating fern *Pteris vittata. Plant Physiology, 141*(4), 1544–1554.

Gao, X., Rodrigues, S. M., Spielman-Sun, E., Lopes, S., Rodrigues, S., Zhang, Y., & Lowry, G. V., (2019). Effect of soil organic matter, soil pH, and moisture content on solubility and dissolution rate of CuO NPs in soil. *Environmental Science & Technology, 53*(9), 4959–4967.

Huang, Q., Liu, Q., Lin, L., Li, F. J., Han, Y., & Song, Z. G., (2018). Reduction of arsenic toxicity in two rice cultivar seedlings by different nanoparticles. *Ecotoxicology and Environmental Safety, 159*, 261–271.

Joo, S. H., & Zhao, D., (2017). Environmental dynamics of metal oxide nanoparticles in heterogeneous systems: A review. *Journal of Hazardous Materials, 322*, 29–47.

Kamiya, T., Islam, R., Duan, G., Uraguchi, S., & Fujiwara, T., (2013). Phosphate deficiency signaling pathway is a target of arsenate and phosphate transporter OsPT1 is involved in As accumulation in shoots of rice. *Soil Science and Plant Nutrition, 59*(4), 580–590.

Kamiya, T., Tanaka, M., Mitani, N., Ma, J. F., Maeshima, M., & Fujiwara, T., (2009). NIP1; 1, an aquaporin homolog, determines the arsenite sensitivity of *Arabidopsis thaliana. Journal of Biological Chemistry, 284*(4), 2114–2120.

Khan, M. R., & Rizvi, T. F., (2017). Application of nanofertilizer and nanopesticides for improvements in crop production and protection. In: *Nanoscience and Plant–Soil Systems* (pp. 405–427). Springer, Cham.

Kotrba, P., Najmanova, J., Macek, T., Ruml, T., & Mackova, M., (2009). Genetically modified plants in phytoremediation of heavy metal and metalloid soil and sediment pollution. *Biotechnology Advances, 27*(6), 799–810.

Lata, S., & Samadder, S. R., (2016). Removal of arsenic from water using nano adsorbents and challenges: A review. *Journal of Environmental Management, 166*, 387–406.

Li, B., Zhou, S., Wei, D., Long, J., Peng, L., Tie, B., & Lei, M., (2019). Mitigating arsenic accumulation in rice (*Oryza sativa* L.) from typical arsenic contaminated paddy soil of southern China using nanostructured α-MnO_2: Pot experiment and field application. *Science of the Total Environment, 650*, 546–556.

Li, R. Y., Ago, Y., Liu, W. J., Mitani, N., Feldmann, J., McGrath, S. P., & Zhao, F. J., (2009). The rice aquaporin Lsi1 mediates uptake of methylated arsenic species. *Plant Physiology, 150*(4), 2071–2080.

Liu, C., Wei, L., Zhang, S., Xu, X., & Li, F., (2014). Effects of nanoscale silica sol foliar application on arsenic uptake, distribution and oxidative damage defense in rice (*Oryza sativa* L.) under arsenic stress. *RSC Advances, 4*(100), 57227–57234.

Liu, J., Dhungana, B., & Cobb, G. P., (2018). Copper oxide nanoparticles and arsenic interact to alter seedling growth of rice (*Oryza sativa japonica*). *Chemosphere, 206*, 330–337.

Liu, J., Simms, M., Song, S., King, R. S., & Cobb, G. P., (2018). Physiological effects of copper oxide nanoparticles and arsenic on the growth and life cycle of rice (*Oryza sativa japonica 'Koshihikari'*). *Environmental Science & Technology, 52*(23), 13728–13737.

Liu, T., Zhu, L., Wang, J., Wang, J., Zhang, J., Sun, X., & Zhang, C., (2015). Biochemical toxicity and DNA damage of imidazolium-based ionic liquid with different anions in soil on *Vicia faba* seedlings. *Scientific Reports, 5*(1), 1–10.

Ma, J. F., & Yamaji, N., (2006). Silicon uptake and accumulation in higher plants. *Trends in Plant Science, 11*(8), 392–397.

Ma, J. F., Yamaji, N., Mitani, N., Xu, X. Y., Su, Y. H., McGrath, S. P., & Zhao, F. J., (2008). Transporters of arsenite in rice and their role in arsenic accumulation in rice grain. *Proceedings of the National Academy of Sciences, 105*(29), 9931–9935.

Ma, X., Sharifan, H., Dou, F., & Sun, W., (2020). Simultaneous reduction of arsenic (As) and cadmium (Cd) accumulation in rice by zinc oxide nanoparticles. *Chemical Engineering Journal, 384*, 123802.

Maity, J. P., Chen, C. Y., Bhattacharya, P., Sharma, R. K., Ahmad, A., Patnaik, S., & Bundschuh, J., (2020). Arsenic removal and mitigation options by advanced application of nano-technological and biological processes. *J. Hazard. Mater., 123885.*

Martinson, C. A., & Reddy, K. J., (2009). Adsorption of arsenic (III) and arsenic (V) by cupric oxide nanoparticles. *Journal of Colloid and Interface Science, 336*(2), 406–411.

Mawia, A. M., Hui, S., Zhou, L., Li, H., Tabassum, J., Lai, C., & Hu, P., (2021). Inorganic arsenic toxicity and alleviation strategies in rice. *Journal of Hazardous Materials, 408*, 124751.

Mukhopadhyay, R., Bhattacharjee, H., & Rosen, B. P., (2014). Aquaglyceroporins: Generalized metalloid channels. *Biochimica et Biophysica Acta (BBA)-General Subjects, 1840*(5), 1583–1591.

Norton, G. J., Pinson, S. R., Alexander, J., Mckay, S., Hansen, H., Duan, G. L., & Price, A. H., (2012). Variation in grain arsenic assessed in a diverse panel of rice (*Oryza sativa*) grown in multiple sites. *New Phytologist, 193*(3), 650–664.

Ochoa, M., Tierra, W., Tupuna-Yerovi, D. S., Guanoluisa, D., Otero, X. L., & Ruales, J., (2020). Assessment of cadmium and lead contamination in rice farming soils and rice (*Oryza sativa* L.) from Guayas province in Ecuador. *Environmental Pollution, 260*, 114050.

Qian, Y., Qin, C., Chen, M., & Lin, S., (2020). Nanotechnology in soil remediation – applications vs. implications. *Ecotoxicology and Environmental Safety, 201*, 110815.

Roduner, E., (2006). Size matters: Why nanomaterials are different. *Chemical Society Reviews, 35*(7), 583–592.

Rossi, L., Sharifan, H., Zhang, W., Schwab, A. P., & Ma, X., (2018). Mutual effects and in planta accumulation of co-existing cerium oxide nanoparticles and cadmium in hydroponically grown soybean (Glycine max (L.) Merr.). *Environmental Science: Nano, 5*(1), 150–157.

Sharma, Y. C., Srivastava, V., Singh, V. K., Kaul, S. N., & Weng, C. H., (2009). Nano-adsorbents for the removal of metallic pollutants from water and wastewater. *Environmental Technology, 30*(6), 583–609.

Shi, S., Wang, T., Chen, Z., Tang, Z., Wu, Z., Salt, D. E., & Zhao, F. J., (2016). OsHAC1; 1 and OsHAC1; 2 function as arsenate reductases and regulate arsenic accumulation. *Plant Physiology, 172*(3), 1708–1719.

Shin, H., Shin, H. S., Dewbre, G. R., & Harrison, M. J., (2004). Phosphate transport in *Arabidopsis*: Pht1; 1 and Pht1; 4 play a major role in phosphate acquisition from both low-and high-phosphate environments. *The Plant Journal, 39*(4), 629–642.

Song, W. Y., Yamaki, T., Yamaji, N., Ko, D., Jung, K. H., Fujii-Kashino, M., & Ma, J. F., (2014). A rice ABC transporter, OsABCC1, reduces arsenic accumulation in the grain. *Proceedings of the National Academy of Sciences, 111*(44), 15699–15704.

Srivastava, N. K., & Majumder, C. B., (2008). Novel biofiltration methods for the treatment of heavy metals from industrial wastewater. *Journal of Hazardous Materials, 151*(1), 1–8.

Sun, S. K., Chen, Y., Che, J., Konishi, N., Tang, Z., Miller, A. J., & Zhao, F. J., (2018). Decreasing arsenic accumulation in rice by overexpressing Os NIP 1; 1 and Os NIP 3; 3 through disrupting arsenite radial transport in roots. *New Phytologist, 219*(2), 641–653.

Syu, C. H., Huang, C. C., Jiang, P. Y., Lee, C. H., & Lee, D. Y., (2015). Arsenic accumulation and speciation in rice grains influenced by arsenic phytotoxicity and rice genotypes grown in arsenic-elevated paddy soils. *Journal of Hazardous Materials, 286*, 179–186.

Verma, P. K., Verma, S., Tripathi, R. D., & Chakrabarty, D., (2020). A rice glutaredoxin regulates the expression of aquaporin genes and modulate root responses to provide arsenic tolerance. *Ecotoxicology and Environmental Safety, 195*, 110471.

Wang, F. Z., Chen, M. X., Yu, L. J., Xie, L. J., Yuan, L. B., Qi, H., & Chen, Q. F., (2017). OsARM1, an R2R3 MYB transcription factor, is involved in regulation of the response to arsenic stress in rice. *Frontiers in Plant Science, 8*, 1868.

Wang, P., Xu, X., Tang, Z., Zhang, W., Huang, X. Y., & Zhao, F. J., (2018a). OsWRKY28 regulates phosphate and arsenate accumulation, root system architecture and fertility in rice. *Frontiers in Plant Science, 9*, 1330.

Wang, P., Yin, N., Cai, X., Du, H., Fu, Y., Geng, Z., & Cui, Y., (2021). Assessment of arsenic distribution, bioaccessibility and speciation in rice utilizing continuous extraction and *in vitro* digestion. *Food Chemistry, 346*, 128969.

Wang, X., Sun, W., & Ma, X., (2019). Differential impacts of copper oxide nanoparticles and copper (II) ions on the uptake and accumulation of arsenic in rice (*Oryza sativa*). *Environmental Pollution, 252*, 967–973.

Wang, X., Sun, W., Zhang, S., Sharifan, H., & Ma, X., (2018b). Elucidating the effects of cerium oxide nanoparticles and zinc oxide nanoparticles on arsenic uptake and speciation in rice (*Oryza sativa*) in a hydroponic system. *Environmental Science & Technology, 52*(17), 10040–10047.

Wang, Z., Von, D. B. A., Kabadi, P. K., Kane, A. B., & Hurt, R. H., (2013). Biological and environmental transformations of copper-based nanomaterials. *ACS Nano, 7*(10), 8715–8727.

Wu, X., Hu, J., Wu, F., Zhang, X., Wang, B., Yang, Y., & Wang, X., (2021). Application of TiO_2 nanoparticles to reduce bioaccumulation of arsenic in rice seedlings (*Oryza sativa* L.): A mechanistic study. *Journal of Hazardous Materials, 405*, 124047.

Wu, Z., Ren, H., McGrath, S. P., Wu, P., & Zhao, F. J., (2011). Investigating the contribution of the phosphate transport pathway to arsenic accumulation in rice. *Plant Physiology, 157*(1), 498–508.

Xu, J., Shi, S., Wang, L., Tang, Z., Lv, T., Zhu, X., & Wu, Z., (2017). OsHAC4 is critical for arsenate tolerance and regulates arsenic accumulation in rice. *New Phytologist, 215*(3), 1090–1101.

Ye, Y., Li, P., Xu, T., Zeng, L., Cheng, D., Yang, M., & Lian, X., (2017). OsPT4 contributes to arsenate uptake and transport in rice. *Frontiers in Plant Science, 8*, 2197.

Younas, A., Yousaf, Z., Riaz, N., Rashid, M., Razzaq, Z., Tanveer, M., & Huang, S., (2020). Role of nanotechnology for enhanced rice production. In: *Nutrient Dynamics for Sustainable Crop Production* (pp. 315–350). Springer, Singapore.

Yruela, I., (2013). Transition metals in plant photosynthesis. *Metallomics, 5*(9), 1090–1109.

Zeng, H., Singh, A., Basak, S., Ulrich, K. U., Sahu, M., Biswas, P., & Giammar, D. E., (2009). Nanoscale size effects on uranium (VI) adsorption to hematite. *Environmental Science & Technology, 43*(5), 1373–1378.

Zhou, S., Peng, L., Lei, M., Pan, Y., & Lan, D., (2015). Control of As soil to rice transfer (*Oryza sativa* L.) with nano-manganese dioxide. *Acta Sci. Circumst., 35*, 855–861.

CHAPTER 9

Microbial-Mediated Measures to Regulate Arsenic in Rice

GAURAV KUMAR SINGH,[1] ANKITA,[2] JITENDER SINGH,[3] and NAVJOT KAUR[4]

[1]*Department of Forensic Science, Sharda School of Allied Health Sciences (SSAHS), Sharda University, Greater Noida, Uttar Pradesh, India*

[2]*Department of Anthropology, Panjab University, India*

[3]*Institute of Pharmaceutical Sciences (IPS Bhaddal), IET Bhaddal Technical Campus, Ropar, Punjab, India*

[4]*Department of Forensic Science, Chandigarh University, Mohali, Punjab, India*

ABSTRACT

Arsenic contamination is a serious hazard to human beings. The deposition of arsenic in soil and water has a critical impact on the health of millions of people. The interference of arsenic is mainly through the soil-water-food chain in human beings. The direct exposure of arsenic to humans is less but chronic exposure may lead to various physiological and psychological effects. The major source of arsenic is found in food and dietary supplements. Rice is a major food ingredient and the uptake of arsenic in rice plants leads to accumulation of arsenic in small quantities which got incorporated into the human food chain. The toxicity and increased appearance in day-to-day life have gained the attention of various researchers

Arsenic in Rice: A Global Menace. Garima Awasthi, Sudhakar Shrivastava, and Mahipal Singh Sankhla, (Eds.)

and healthcare professionals for the mitigation of toxic metals (loids). Bioremediation is a proving technique that uses living microorganisms for the removal of toxic metalloids. Bioremediation and its techniques have been included in this chapter which mainly focuses on the elimination of arsenic from soil and groundwater. In this chapter, the transportation of arsenic in plants majorly in rice, the role of microbes in remediation, and its mechanism have been discussed.

9.1 INTRODUCTION

Arsenic is a major metallic pollutant highly incorporated into the food chain. Arsenic in the environment is widely spread in various parts of the world majorly in Southeast Asia, Europe, and China (Srivastava et al., 2016b). The toxicity of arsenic depends on its several oxidation states. Arsenate [As(V)], arsenite [As(III)], and elemental As are mostly found in the environment. Among them, arsenite [As(III)] is the most toxic and hazardous state. Due to climatic changes and industrialization, increases in arsenic levels are seen, leading to contamination of soil and biomagnification in the soil-plant-human system. Arsenic has major applications in the manufacturing of pesticides and insecticides. However, some species have gained resistance by developing their respiratory metabolism (Yang & Rosen, 2016). Arsenic causes major health issues in humans, such as cancers and genotoxicity (World Health Organization, 2004). According to WHO guidelines, 10 μg/L is the permissible limit of Arsenic (World Health Organization, 2004). Although, As is not only harmful to human beings but also its presence in soil alters various crucial properties. Majorly, being a metallic element, it changes the physicochemical properties of soil which ultimately affects the yielding productivity of crops. Rice is widely used as a food and its cultivation is heavily taken up. However, the water requirement for the cultivation of rice is more; the chances of contamination with arsenic also increase. As(III) and As(V) are the dominant states that contaminate the soil.

9.1.1 *BIOREMEDIATION OF ARSENIC*

The process comprises the utilization of plants or microorganisms to eliminate the contaminants or toxins present in the environment (soil, groundwater). The *in-situ* process of bioremediation is better adapted in case there is any occurrence of diffused contamination for medicines of many regions. That treated land area opens up for more secure uses. Arsenic being a metalloid, rather than various normal pollutions, can't be changed over to CO_2 and dispersed into the environment. Arsenic can exist in its four oxidation states (+5, +3, 0, and –3). The most common commonly found states are "Arsenate, As(V)" and "Arsenite, As(III)" but "Arsenite" is more toxic than "Arsenate" (Kamiya et al., 2013; Wu et al., 2011). The conceivable method of bioremediation would be oxidation of the more destructive As(+3) to the less harmful As(+5), methylation of inorganic species, and extraction (take-up) by plants and a short time later discarding the incomparable As plant biomass (Wu et al., 2011; Catarecha et al., 2007; Zhang et al., 2017).

9.2 BIOREMEDIATION PRACTICE TO MITIGATE ARSENIC IN RICE

Among the list of toxic elements, arsenic (As) is positioned at number 5 and classified as a Class I carcinogenic substance; also known as – "*king of poison*." Arsenic is an unnecessary element that is harmful to crop production in agricultural fields as it is reducing plant yield which is a matter of great environmental concern worldwide (Mendoza-Cózatl et al., 2011). Various anthropogenic activities lead to the deposition of arsenic in the soil, destructing the properties of soil and ultimately decreasing agricultural production. Bioremediation is defined as the treatability technology utilizing biological activities for minimizing or degrading the toxic heavy metals from the soil commonly using microorganisms like bacteria, fungi, or some plant species which degrade the toxic elements through their metabolic pathways (Alka et al., 2020; Hossain, 2012; Apel & Hirt, 2004).

Plants take up the essential nutrients and minerals from soil and water through transporters. These transporters readily absorb the essential components along with nonessential elements. Inorganic forms of arsenic, i.e., Arsenate [As(V)] and Arsenite [As(III)] are the most toxic. Both differ in their toxicity and mode of transport as As(V) transports through phosphate transporters (Chauhan et al., 2017). There are many transporters that help As to get into cells as OsPHT1;1 (Kamiya et al., 2013), OsPHT1;8 (Wu et al., 2011) in rice, and AtPHT1;1, AtPHT1;4, AtPHT1;5, AtPHT1;7,

AtPHT1;8, AtPHT1;9. Among these, the OsPHT1; 8 transporter uptakes the As(V) in rice. Apart from this transporter, PHT proteins help in taking As(V) from root to top shoot (Wu et al., 2011; Catarecha et al., 2007; Mendoza-Cózatl et al., 2011). The methylated arsenic (Jia et al., 2013) elements and As(III) are taken up by aquaglyceroporins, mainly through nodulin 26-like intrinsic protein (NIP) (Alka et al., 2020). OsNIP2; 1 (Lsi1) transports Silicic acid along with As(III). More specifically, OsNIP1;1, OsNIP2;2 (OsLsi6), OsNIP3;1, OsNIP3;2, OsNIP3;3 are the major transporters of As(III) in rice (Ma et al., 2008). A recent study stated that OsNIP3;3 is the predominant transporter in rice via lateral roots (Chen et al., 2017). The bioremediation approach is likely to be more sustainable which utilizes both the plant as well as the microorganisms having the resistance to the toxic heavy metals and the capability of surviving in the contaminated soil. The bioremediation approach has various advantages mainly including sustainability, decreased cost, ensuring worker safety, etc. The bioremediation approach includes phytoremediation, and microbial bioremediation (Hou et al., 2020). Usually, the metal contaminants introduced due to human activities are present in the environment in various matrixes such as soil, water, and sediments. There is a need for remediation of metal contaminants because these are non-biodegradable in nature. Therefore, bioremediation is an eco-friendly method for degrading metal pollutants. There are various ways of bioremediation which have been discussed further in this chapter (Kumar, Yadav, & Ramanathan, 2020). Arsenic can be remediated by utilizing both conventional as well as bioremediation methods. But the conventional methods are not eco-friendly like the bioremediation methods. The different types of conventional and bioremediation methods have been represented in Figure 9.1 (Sher & Rehman, 2019). Rice is among the main staple food required for nutritional protection but the crop is getting destroyed due to deposition of the various contaminants such as the toxic metalloids – arsenic, cadmium, mercury, and lead also known as the '*most problematic heavy metals*' or '*toxic heavy metals*' (Rahman & Singh, 2020) present in the environment due to the natural and anthropogenic activities. Accumulation of these toxic metals is not harmful to plants but also to animals and humans. The production of contaminated crops poses a public health risk globally, and action should be taken for the mitigation of this problem (Saxena, Kumar, & Shah, 2020).

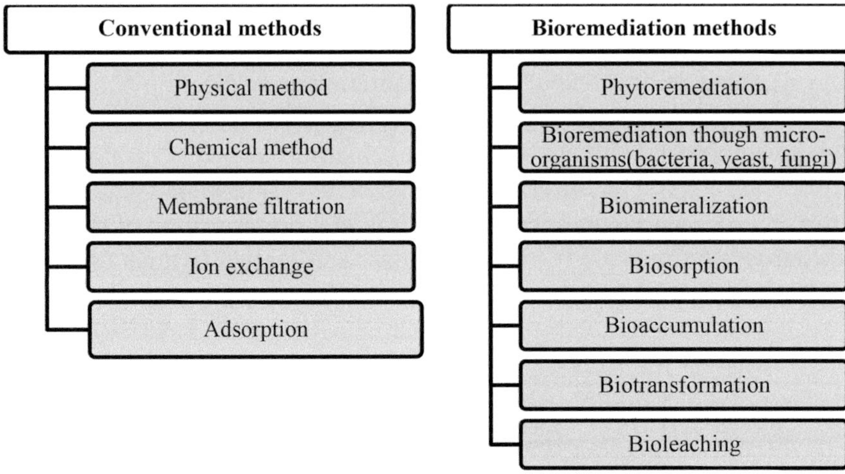

FIGURE 9.1 Representing the methods for arsenic remediation.

9.3 MECHANISMS OF MICROBIAL MEDIATED AS BIOTRANSFORMATION

9.3.1 MICROBE-BASED MECHANISM

Microorganisms play a significant role in the remediation of arsenic from the paddy fields. By influencing the biotransformation, bioaccumulation, and bioavailability of arsenic in the rice ecosystem (Khanam et al., 2020). The basis of the bioremediation depends on the microbes' interaction with the heavy metals. Microorganisms such as bacteria, yeast, fungi, algae, lichens, protozoan, and archaebacteria have the capability of moderating some sort of heavy metal interactions. The toxic heavy metals even at lower concentrations pose toxic effects on non-resistant microorganisms by copying the required elements and hindering the biological working and production of oxygen species and inhibiting the biomolecules such as DNA, RNA, and protein (Rahman & Singh, 2020). Whereas arsenic-resistant microorganisms can even survive under the influence of higher concentrations of Arsenic which could be lethal for other organisms.

Some species of As-tolerant bacterial strains as shown in Figure 9.2 can decrease the deposition of arsenic by 30–40% in rice grains when grown in the As contaminated rice fields. Therefore, the microbes-based remediation method has been found to the detoxification phenomenon in the rice ecosystems. A few of the microorganisms degrading the arsenic are autotrophs and heterotrophs (Saxena, Kumar, & Shah, 2020). Autotrophic microbes utilize the As(III) for respiration as an electron donor while the heterotrophic microbes consume the organic carbon for their growth and development. Whereas some bacterial strains such as *Paracoccus* sp. oxidize the As(III) as an electron acceptor by using either NO_3^- or O_2. Other microbe-based practices for minimizing the deposition of arsenic in rice agroecosystems are methylation and volatilization (Kumarathilaka et al., 2020) (Figure 9.3).

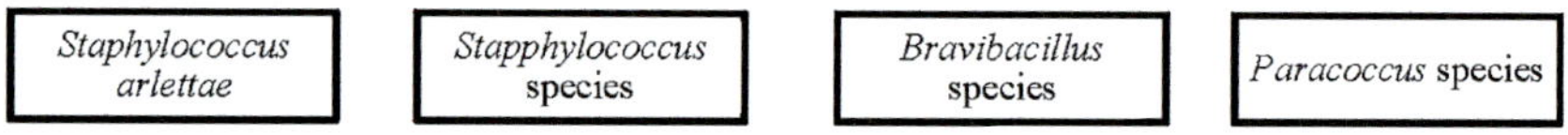

FIGURE 9.2 Some examples of arsenic (As) tolerant bacterial strains.

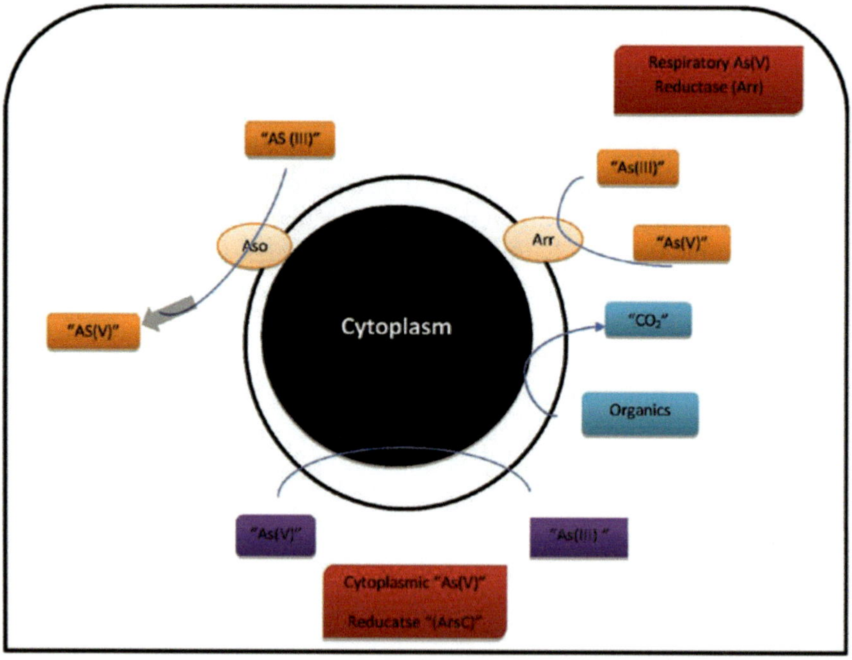

FIGURE 9.3 Inorganic As species showing biochemical transformation by microbial cells.

9.3.2 PHYTOEXTRACTION: ARSENIC UPTAKE ENHANCEMENT

The mechanism of Arsenic movement and uptake proved significant with the help of hyperaccumulators such as *P. vittata*. The species *P. vittata* has been observed as a significant carrier for the uptake of As(V) from the soil. This has been preferred over *A. thaliana* as As(V) carrier by extracting a huge amount of arsenic from debased grounds or waters (Ditusa et al., 2016). *P. vittate* has a higher tendency for arsenate extraction than non-hyperaccumulator plants. The process involves Arsenic Uptake, Arsenic transport, and detoxification.

9.3.3 BIOCHAR EFFECTS ON "SOIL-PLANT-EARTHWORM-BACTERIA SYSTEM" IN METALLOID-POLLUTED SOIL

To put it plainly, the expansion of biochar to tainted soil expanded the convergence of accessible arsenic and diminished accessible cadmium in soils. In any case, arsenic, and cadmium takes-up by the plant were diminished after biochar corrections. Nightcrawlers as the major macrofauna in soils can be applied as a bioindicator for soil pollution, and utilized in the remediation of metal(loid) polluted soils (Biederman & Harpole, 2013). In addition, the huge contrast between the bacterial networks of the dirt and night crawler stomach was noticed, mirroring the progress from the encompassing soil to the worm's extraordinary stomach living space. Contrasted and the control, RSBC revisions prompted huge changes in the night crawler stomach microbes, reasonably because of more natural mixtures assets in soils altered with RSBC. The use of biochar revisions to debased soil stays disputable, and extra examinations should be performed to explore the impacts of biochar application on accessible metal(loid) and the environmental framework including plants, creatures, and microbiota. The night crawlers could be added to speed up the metal(loid) expulsion from the defilement soils (Dendooven et al., 2013).

9.3.4 THE EFFECT ON ARSENIC ACCUMULATION IN RICE (ORYZA SATIVA L.) DUE TO RHIZOSPHERE BACTERIAL COMMUNITY

The physicochemical properties of soil, rice cultivars, and soil microbial exercises directly affect the accumulation of metalloids in rice plants. The sterilization of soil does not affect the physicochemical properties of soil along with the bioavailability of metalloids in soil. In any case, the interference of Arsenic in rice plants is less in the sterilized soil than in the untreated one that held the local bacterial neighborhood. The microbial neighborhood of the rhizosphere plays a great role in the total of profound metalloids in plants (Dong, Feng, & Wang, 2016).

9.4 BIOREMEDIATION TECHNIQUE IS CATEGORIZED INTO TWO PARTS

9.4.1 IN SITU

In situ – the most appropriate technique utilized for soil and groundwater with a minimum disturbance at the actual polluted site without any mining or dumping.

9.4.2 EX SITU

Ex situ utilized for soil and groundwater at a site that has been detached via mining and pumping from the actual contaminated site to another site via excavation and pumping for the degradation of contaminants using the various approaches of bioremediation (Saxena, Kumar, & Shah, 2020; Najam, Shah, & Rahman, 2021).

There is not any particular absorption pathway for arsenic in the microbes and don't have any significant role in nutrient supply and metabolism in the cytoplasm. Due to the similarity in the chemical structure, Arsenic enters into the cells of microorganisms via the phosphate transporters. Microorganism interacts with the arsenic physically via both intra and extra-cellular and tries to remove the arsenic from either binding into the cells or excluding it from the cells. The various mechanisms of microbial interactions with arsenic have been mentioned in further sections (Figure 9.4) (Hare, Chowdhary, & Singh, 2020).

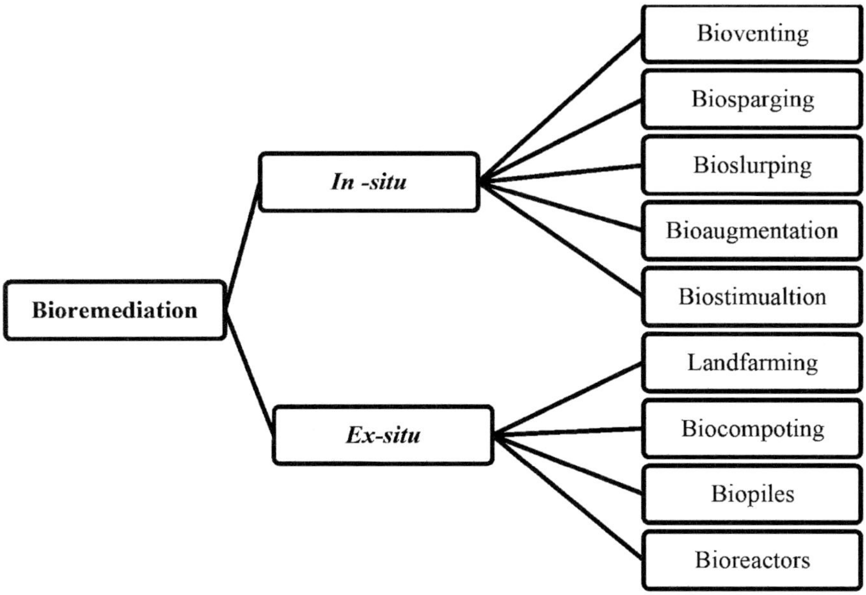

FIGURE 9.4 Techniques of bioremediation.

9.5 MECHANISMS OF MICROBIAL BIOREMEDIATION

The bioremediation approach involves the utilization of microorganisms like bacteria which have the ability to detox the contaminants in soil and water. These bacteria are capable of cleaning up oil spills, petroleum elements, and heavy metals from contaminated soil and water or other environments (Figure 9.5).

9.5.1 BIOACCUMULATION

Bioaccumulation and biosorption are two different concepts, differing in their mechanisms. Bioaccumulation is a complex metabolically active process in which microorganisms uptakes heavy metals with the help of an importer complex which creates a pathway through the lipoprotein bilayer. Later the uptaken heavy metals are detoxified by the peptide glands and proteins (Dixit et al., 2015; Zhao et al., 2009). Proteins in the cytosol and lipid membranes help in decreasing cell viability which expresses the heterologous import storage proteins, efficiently helping microbes to

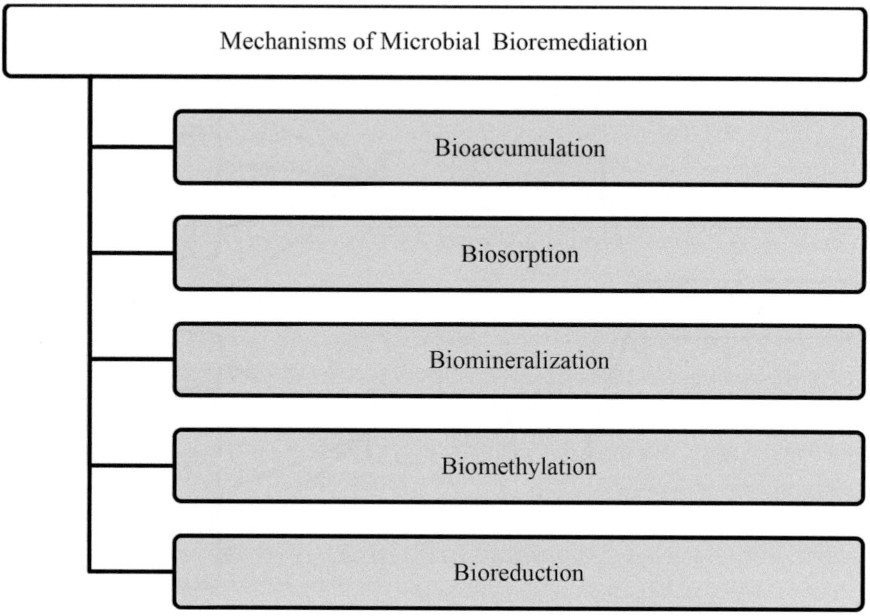

FIGURE 9.5 Mechanisms of microbial-based arsenic bioremediation.

uptake HMs. This bio-accumulative capacity is based on μmol_x or mg_x per $g_{dry\ weight}$, where x is the HM and these values are mentioned for cadmium (Cd), cobalt (Co), copper (Cu), mercury (Hg), nickel (Ni), uranium (U), and the metalloid arsenic (As^{3+}, As^{4+}).

9.5.2 BIOSORPTION

Biosorption is a metabolically passive process of biological materials having the ability to accumulate HMs from contaminated water and soil. In biosorption, microorganisms through the spontaneous physicochemical pathways uptakes the HMs in their cellular space. Various physicochemical factors like ionic strength, the chemistry of metal ions, metal concentration, temperature, and contact time are responsible. Biologic factors like the cell wall composition of microbes are also another factor as for the uptake of HMs in cellular space, the importer protein must be present in the cell wall (Saba, 2018; Shukla, Sarim, & Singh, 2020).

9.5.3 BIOMINERALIZATION

Biomineralization is a detoxifying biological process in which microorganisms form the minerals which help out in the removal of heavy metals. These minerals are the byproduct of microorganisms which are the results of a number of chemical reactions. Oxides, sulfides, phosphates, and oxalates are some of the biominerals expediated by microbes. Biomineralization is an active process in an environment in which microorganisms produce metabolic products which are reacting with compounds to give minerals. There are two types of biomineralization, "biologically induced mineralization (BIM)" and "biologically controlled mineralization (BCM)." In the first one, the extra-cellular mineral production is regulated as a result of geochemical reactions. In this, microbes have no control over the production of minerals. While, in BCM, the reactions and production of minerals are controlled by microorganisms intracellularly or epicellular. Microbes can generate nearly 64 different types of minerals listing as iron and manganese oxides, sulfide minerals, phosphorites, and many more. Calcite is one biomineralized byproduct that can actively adsorb metalloid compounds in its crystal structure (Knoll, 2003; Webb et al., 2006; Pan, 2009; DiBenedetto et al., 2006).

9.5.4 BIOMETHYLATION

Biomethylation is an efficient detoxification process that reduces mobility, volatility, and toxicity. Here, microbial decomposition acts as an indirect mechanism that mobilized the metal impurities in the aqueous phase and soil. Some microorganisms are known to produce dissolved organic matter with releasing of heavy metals. They choose the chelation mechanism in which the chelating of metal ions with carboxylic acids and amino acids is regulated. Many different mechanisms are identified by researchers for metal immobilization bioconversion and inter/intracellular precipitation. With the help of immobilization, heavy metals can be removed efficiently, i.e., As(III) and As(V) are immobilized by bacterial oxidation (Kapahi & Sachdeva, 2019).

9.5.5 BIOREDUCTION

Bioreduction is a process mainly based on electron acceptance and donation which leads to the gain of energy in the absence of oxygen. This results in a decrease in the solubility of metalloid compounds in the environment. In the deficiency of oxygen, microbes accept the electrons resulting in a reduction of nitrates and sulfates. In the end, the carbon dioxide is reduced to methane. In the presence of organic matter, nitrate-metal reduction and sulfate-metal reduction occur consequently. Here, the factors affecting are pH and temperature. Mostly, by this bioremediation technique, Mn (IV), Fe(III), and U(VI) to U(IV) are reduced efficiently (Laura, Katherine, & Jonathan, 2014).

9.5.6 ROLE OF MICROBES IN BIOREMEDIATION

9.5.6.1 BIOREMEDIATION BY BACTERIA

The application of soil bacteria for the bioremediation of Arsenic in the rice agroecosystems has gained attention over the last few years due to the advancements in the biochemical and molecular field after the recognition of beneficial microbes having a significant role in the bioremediation process (Dong et al., 2016; Mani, & Kumar, 2014). The state of arsenic actually depends on the environment in which it resides. This develops the resistance capability in the microbes to decrease the harmful effect of active extrusion, intracellular sequestration, chelation, or precipitation. There are some microbes that are efficient to use the arsenic compound in their metabolism for the anaerobic respiration process (Satyapal, Rani, Kumar, & Kumar, 2016). The bacteria *Corynebacterium glutamicum* is one of the most arsenic-resistant microorganisms. This is used for the production of amino acids and nucleotides in industry. In bacteria involving remediation, arsenic regulation is performed by two mechanisms. One way is using arsenic carrier protein in which the energy is supplied by the membrane potential of the cell and another mechanism is arsenite translocating ATPase (Mateos, Ordóñez, Letek, & Gil, 2006) (Figure 9.6).

9.5.6.2 BIOREMEDIATION USING FUNGI

Fungi play a crucial role in the process of bioremediation, possessing unique metabolic capacity and vigorous structure. The bioremediation process using fungi is known as *Mycoremediation* and is the most sustainable and green route of metalloid regulation from the environment or contaminated locations (Deshmukh, Khardenavis, & Purohit, 2016).

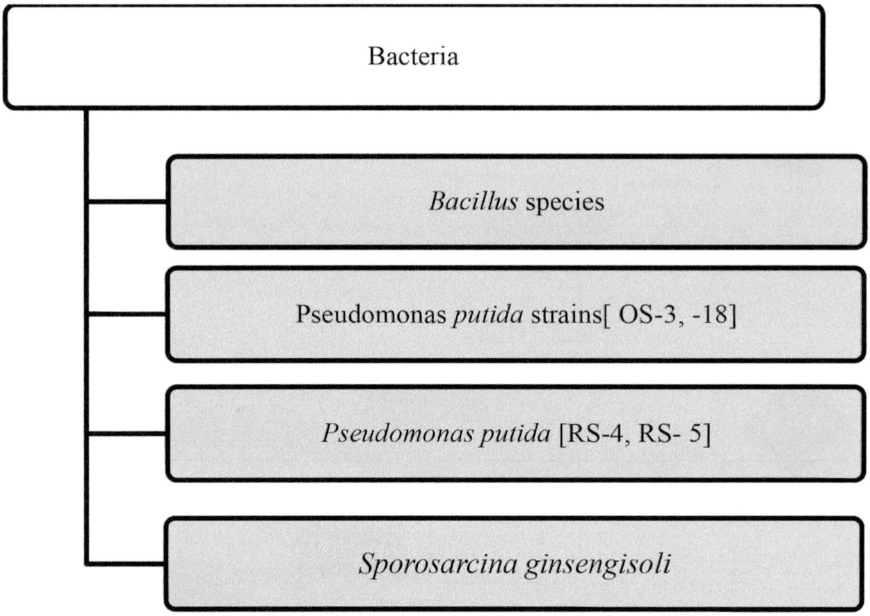

FIGURE 9.6 Examples of bacteria used for bioremediation.

Mycorrhiza is the symbiotic association of most terrestrial plants, which enhances the transportation of nutrients and resistance to biotic as well as abiotic stress. The effects of arsenic have been examined when introducing the *Arbuscular mycorrhiza* (AM) fungus, i.e., *Glomumosseae on* a lentil *Lens culinaris* which is an edible legume plant. After subjecting the fungus to the plant, it was found that the height of the plant, number of leaves, pod height, biomass, etc., increased significantly due to the *mycorrhizal* and also minimized the concentration of arsenic in the root and shoot system of the plant. The results of the study (Ahmed, Killham, & Alexander, 2006) suggested that growing the lentils with the *Arbuscular mycorrhiza* will lead to a more uptake of nutrients, mainly phosphorus, and minimize arsenic contamination (Hossain, 2012) (Figure 9.7).

9.5.6.3 BIOREMEDIATION USING ALGAE

Numerous studies have reported the use of algae in bioremediation because algae are capable of hyperaccumulating the arsenic from the water and

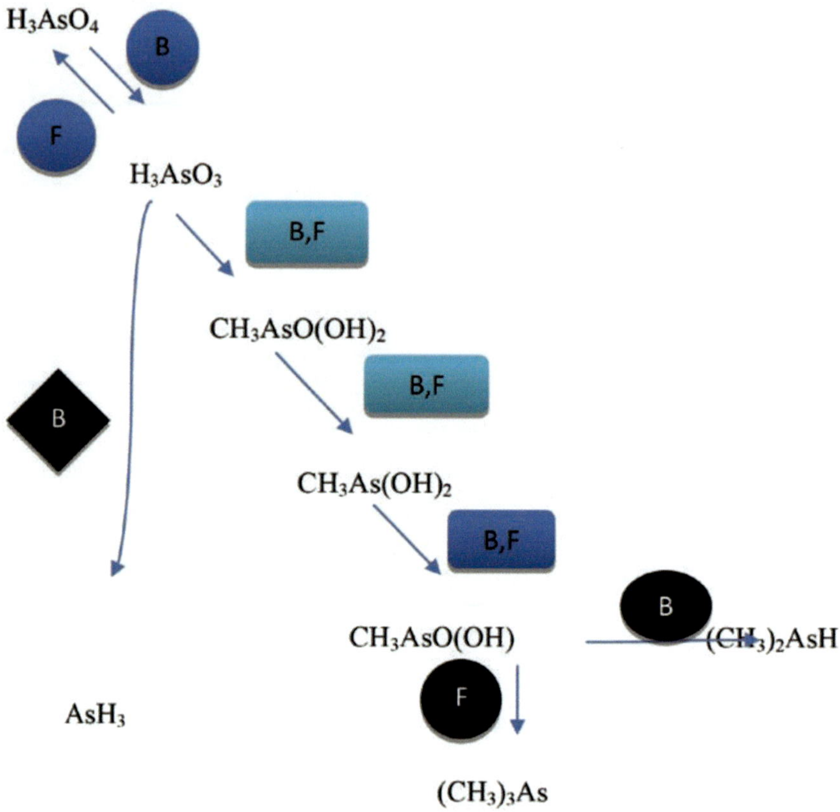

FIGURE 9.7 Summary of observed microbial interactions with arsenic (As) species performed by bacteria (B) and fungi (F).

this specialty of algae is utilized for the degradation of arsenic from the arsenic-contaminated rice ecosystem. Some examples of algae utilized for bioremediation include – species of *Chlorella, Chlamydomonas, Gloeotrichia, Hydrodictylon, Lyngbya, Navicula, Oscillatoria, Oedogonium, Rhizoclonium*, etc. (Pan, 2009; Hossain, 2012).

9.5.6.4 BIOREMEDIATION USING YEAST

High-affinity phosphate transporters [Pho 84] are responsible for the uptake of As in the yeast cells and there is also a role of glucose permeases in this process. The clusters of genes that are encoding for the tolerance of arsenic in the yeast mainly comprise ACR1, ACR2, and ACR3 which are responsible for the sensing, reduction, and efflux of arsenic (Saba, 2018). The yeast boosts the photosynthetic pigments in rice plants in control conditions that get affected by arsenic presence. The genetically engineered (GE) yeast suppresses arsenic accumulation and also significantly affects the morphology of rice plants (Verma, Verma, & Chakrabarty, 2019).

9.6 CONCLUSION

Urbanization and natural calamities are major sources of heavy metal contamination in groundwater and soil. For the removal of toxic metalloid compounds incorporated in soil and water, there are a number of techniques in the literature. Microorganisms are the key factors in the environment that plays a crucial role. They have developed some reflexes against these heavy metals for survival in contaminated soil and water. Bioremediation is one of the promising techniques in which microorganisms' biological processes are employed for the removal of toxic metalloids. There are some specific bacteria that accumulate the heavy metals and undergo various biological processes and release the oxidized or reduced compound which is less harmful to nature. Microorganisms have demonstrated the potential to filter, concentrate, exclude, or biotransform a series of rising contaminants. The treatment methods mainly comprise bio-adsorption, bio-uptake, bio-degradation, etc. Arsenic contamination is a serious challenge worldwide affecting plants, animals including human beings. Plants and microbes should be merged to conquer the problem of Arsenic contamination in rice or environmental contamination of arsenic. This particular approach of bioremediation is found to be as eco-friendly, low cost, less environmental destruction, and a possible solution to deal with Arsenic contamination. Further, there is a need to analyze the impacts of environmental conditions on plant–microorganism-based interactions in the contaminated soils. Microorganisms, particularly bacteria of different genera are found as great bio-remediators for the lessening of toxic heavy metals under various bioprocess as explained in this chapter. There are also other techniques of genetic engineering and nanotechnology developing in the present time to improve the capability of bio-remediators. The benefits and the risks related to the bio-processes should also be considered after the field trials. Therefore, the researchers need to focus on these directions to implement the advanced techniques at a large scale.

9.7 FUTURE RECOMMENDATIONS

There is yet the need for development in this sector as there are very few microorganisms that have evolved processes that can reduce the severity and toxicity of metals. Recombinant technology is the only technique through which the incorporation of a specific gene in microorganisms can be done. With the help of these microbes, the removal of contaminants can be performed. Also, researchers need to pay attention to useful microbes which increase fertility and maintain the nutritional value of soil.

KEYWORDS

- **arsenate**
- **arsenic**
- **arsenite**
- **microbial**
- **microorganisms**
- **nutritional value**
- **rice**
- **toxic heavy metals**
- **transporters**

REFERENCES

Ahmed, F. S., Killham, K., & Alexander, I., (2006). Influences of *Arbuscular mycorrhizal* fungus *Glomus mosseae* on growth and nutrition of lentil irrigated with arsenic contaminated water. *Plant and Soil, 283*(1), 33–41.

Alka, S., Shahir, S., Ibrahim, N., Chai, T. T., Bahari, Z. M., & AbdManan, F., (2020). The role of plant growth promoting bacteria on arsenic removal: A of existing perspectives. *Environmental Technology & Innovation, 17*, 100602. doi: 10.1016/j.eti.2020.100602. Available at: https://pubag.nal.usda.gov/catalog/6815995 (accessed on 29 March 2023).

Apel, K., & Hirt, H., (2004). Reactive oxygen species: Metabolism, oxidative stress, and signal transduction. *Annu. Rev. Plant Biol., 55*, 373–399. doi: 10.1146/ annurev. arplant.55.031903.141701.

Biederman, L. A., & Harpole, W. S., (2013). Biochar and its effects on plant productivity and nutrient cycling: A meta-analysis. *GCB. Bioenergy, 5*, 202–214.

Catarecha, P., Segura, M. D., Franco-Zorrilla, J. M., García-Ponce, B., Lanza, M., Solano, R., et al., (2007). A mutant of the *Arabidopsis* phosphate transporter PHT1; 1 displays enhanced arsenic accumulation. *Plant Cell, 19*, 1123–1133.

Catarecha, P., Segura, M. D., Franco-Zorrilla, J. M., García-Ponce, B., Lanza, M., Solano, R., et al., (2007). A mutant of the *Arabidopsis* phosphate transporter PHT1; 1 displays enhanced arsenic accumulation. *Plant Cell, 19*, 1123–1133. doi:10.1105/tpc.106.041871 Available at: https://www.ncbi.nlm.nih.gov/pmc/articles/PMC1867351/ (accessed on 29 March 2023).

Chauhan, R., Awasthi, S., Tripathi, P., Mishra, S., Dwivedi, S., Niranjan, A., et al., (2017). Selenite modulates the level of phenolics and nutrient element to alleviate the toxicity of arsenite in rice (*Oryza sativa* L.). *Ecotox. Environ. Saf., 138*, 47–55. doi: 0.1016/j.ecoenv.2016.11.015.

Chen, Y., Sun, S. K., Tang, Z., Liu, G., Moore, K. L., Maathuis, F. J. M., et al., (2017). The nodulin 26-like intrinsic membrane protein OsNIP3;2 is involved in arsenite uptake by lateral roots in rice. *J. Exp. Bot., 12*, e0173681. doi: 10.1093/jxb/erx165. Available at: https://www.ncbi.nlm.nih.gov/pmc/articles/PMC2481375/ (accessed on 29 March 2023).

Dendooven, L., Peres, G., Tondoh, J. E., Cluzeau, D., & Brun, J. J., (2013). A review of earthworm impact on soil function and ecosystem services. *Eur. J. Soil. Sci., 64*, 161–182.

Deshmukh, R., Khardenavis, A. A., & Purohit, H. J., (2016). Diverse metabolic capacities of fungi for bioremediation. *Indian Journal of Microbiology, 56*(3), 247–264. https://doi.org/10.1007/s12088-016-0584-6.

DiBenedetto, F., Costagliola, P., Benvenuti, M., Lattanzi, P., Romanelli, M., & Tanelli, G., (2006). Arsenic incorporation in natural calcite lattice: Evidence from electron spin echo spectroscopy. *Earth Planet. Sci. Lett., 246*, 458–465. doi: 10.1016/j.epsl.2006.03.047. Available at: https://www.sciencedirect.com/science/article/abs/pii/S0012821X06002822 (accessed on 29 March 2023).

Ditusa, S. F., Fontenot, E. B., Wallace, R. W., Silvers, M. A., Steele, T. N., Elnagar, A. H., et al., (2016). A member of the phosphate transporter 1 (Pht1) family from the arsenic-hyperaccumulating fern *Pteris vittata* is a high-affinity arsenate transporter. *The New Phytologist, 209,* 762–772. doi: 10.1111/ nph.13472.

Dixit, G., Singh, A. P., Kumar, A., Singh, P. K., Kumar, S., Dwivedi, S., et al., (2015). Sulfur mediated reduction of arsenic toxicity involves efficient thiol metabolism and the antioxidant defense system in rice. *J. Hazard. Mater., 298*, 241–251. doi: 10.1016/j.jhazmat.2015.06.008.

Dong, M. F., Feng, R. W., & Wang, R. G., (2016). Inoculation of Fe/Mn-oxidizing bacteria enhances Fe/Mn plaque formation and reduces Cd and As accumulation in rice plant tissues. *Plant Soil, 404*, 75–83. https://doi.org/10.1007/s11104-016-2829-x.

Hare, V., Chowdhary, P., & Singh, A. K., (2020). Arsenic toxicity: Adverse effect and recent advance in microbes mediated bioremediation. *Microorganisms for Sustainable Environment and Health,* 53–80. Available at: https://www.researchgate.net/publication/343055279_Arsenic_toxicity_adverse_effect_and_recent_advance_in_microbes_mediated_bioremediation (accessed on 29 March 2023).

Hossain, M. Z., (2012). Bioremediation of arsenic: Prospects and limitations in the agriculture of Bangladesh. *International Journal of Economic and Environment Geology, 3*(1), 6–14.
Hossain, M. Z., (2012). Bioremediation of arsenic: Prospects and limitations in the agriculture of Bangladesh. *International Journal of Economic and Environment Geology, 3*(1), 6–14.
Hou, D., O'Connor, D., Igalavithana, A. D., Alessi, D. S., Luo, J., Tsang, D. C., & Ok, Y. S., (2020). Metal contamination and bioremediation of agricultural soils for food safety and sustainability. *Nature Reviews Earth & Environment, 1*(7), 366–381.

Jia, Y., Huang, H., Zhong, M., Wang, F. H., Zhang, L. M., & Zhu, Y. G., (2013). Microbial arsenic methylation in soil and rice rhizosphere. *Environmental Science & Technology, 47*(7), 3141–3148.

Kamiya, T., Islam, R., Duan, G., Uraguchi, S., & Fujiwara, T., (2013). Phosphate deficiency signaling pathway is a target of arsenate and phosphate transporter OsPT1 is involved in as accumulation in shoots of rice. *Soil Sci. Plant Nutr., 59*, 580–590. doi: 10.1080/00380768.2013.804390.

Kapahi, M., & Sachdeva, S., (2019). Bioremediation options for heavy metal pollution. *Journal of Health & Pollution, 9*(24), 191203. https://doi.org/10.5696/2156-9614-9.24.191203.

Khanam, R., Kumar, A., Nayak, A. K., Shahid, M., Tripathi, R., Vijayakumar, S., &Pathak, H., (2020). Metal (loid) s (As, Hg, Se, Pb and Cd) in paddy soil: Bioavailability and potential risk to human health. *Science of the Total Environment, 699*, 134330.

Knoll, A., (2003). 'Biomineralization and evolutionary history.' *Reviews in Mineralogy and Geochemistry, 54*, 329–356. doi: 10.2113/0540329. Available at: https://www.semanticscholar.org/paper/Biomineralization-and-Evolutionary-History-Knoll/1cbd47d60407fd534ebd4b566d891ccf328f47f6 (accessed on 29 March 2023).

Kumar, M., Yadav, A., & Ramanathan, A. L., (2020). Arsenic contamination in environment, ecotoxicological and health effects, and bioremediation strategies for its detoxification. In: *Bioremediation of Industrial Waste for Environmental Safety* (pp. 245–264). Springer, Singapore.

Kumarathilaka, P., Seneweera, S., Ok, Y. S., Meharg, A. A., & Bundschuh, J., (2020). Mitigation of arsenic accumulation in rice: An agronomical, physico-chemical, and biological approach–A critical review. *Critical Reviews in Environmental Science and Technology, 50*(1), 31–71.

Laura, N., Katherine, M., & Jonathan, R. L., (2014). The biogeochemistry and bioremediation of uranium and other priority radionuclides. *Chemical Geology, 363*, 164–184. ISSN 0009-2541. https://doi.org/10.1016/j.chemgeo.2013.10.034.

Ma, J. F., Yamaji, N., Mitani, N., Xu, X. Y., Su, Y. H., McGrath, S. P., et al., (2008). Transporters of arsenite in rice and their role in arsenic accumulation in rice grain. *Proc. Natl. Acad. Sci. U.S.A., 105*, 9931–9935. doi: 10.1073/pnas. 0802361105.

Mani, D., & Kumar, C., (2014). Biotechnological advances in bioremediation of heavy metals contaminated ecosystems: An overview with special reference to phytoremediation. *International Journal of Environmental Science and Technology, 11*(3), 843–872.

Mateos, L. M., Ordóñez, E., Letek, M., & Gil, J. A., (2006). *Corynebacterium glutamicum* as a model bacterium for the bioremediation of arsenic. *International Microbiology, 9*(3), 207–215.

Mendoza-Cózatl, D. G., Jobe, T. O., Hauser, F., & Schroeder, J. I., (2011). Long-distance transport, vacuolar sequestration, tolerance, and transcriptional responses induced by cadmium and arsenic. *Curr. Opin. Plant Biol., 14*, 554–562. doi: 10.1016/j.pbi.2011.07.004.
Mendoza-Cózatl, D. G., Jobe, T. O., Hauser, F., & Schroeder, J. I., (2011). Long-distance transport, vacuolar sequestration, tolerance, and transcriptional responses induced by cadmium and arsenic. *Curr. Opin. Plant Biol., 14*, 554–562. doi: 10.1016/j.pbi.2011.07.004.
Najam, T., Shah, S. S. A., &Rahman, M. M., (2021). Water-stable metal–organic framework for environmental remediation. In: *Bioremediation for Environmental Sustainability* (pp. 585–621). Elsevier. Available at-https://www.researchgate.net/publication/345759011_Chapter_24_-_Water-stable_metal-organic_framework_for_environmental_remediation (accessed on 29 March 2023).

Pan, X. L., (2009). Micrologically induced carbonate precipitation as a promising way to in situ immobilize heavy metals in groundwater and sediment. *Res. J. Chem. Environ., 13*, 3, 4. doi: 10.1186/s40064-016-1869-2. Available at: https://www.ncbi.nlm.nih.gov/pmc/articles/PMC4771655/ (accessed on 29 March 2023).

Rahman, Z., & Singh, V. P., (2020). Bioremediation of toxic heavy metals (THMs) contaminated sites: Concepts, applications and challenges. *Environmental Science and Pollution Research, 27*(22), 27563–27581.

Saba, S., (2018). In: Jan, D., & Branislav, V., (eds.), *Biosorption of Heavy Metals, Biosorption.* IntechOpen, doi: 10.5772/intechopen.72099. Available at: https://www.intechopen.com/chapters/58112 (accessed on 29 March 2023).

Satyapal, G. K., Rani, S., Kumar, M., & Kumar, N., (2016). Potential role of arsenic resistant bacteria in bioremediation: Current status and future prospects. *J. Microb. Biochem. Technol., 8*, 256–258. doi: 10.4172/1948-5948.1000294.

Saxena, G., Kumar, V., & Shah, M. P., (2020). *Bioremediation for Environmental Sustainability: Toxicity, Mechanisms of Contaminants Degradation, Detoxification and Challenges*. Elsevier. Available at: https://www.elsevier.com/books/bioremediation-for-environmental-sustainability/saxena/978-0-12-820524-2 (accessed on 29 March 2023).

Sher, S., & Rehman, A., (2019). Use of heavy metals resistant bacteria—A strategy for arsenic bioremediation. *Applied Microbiology and Biotechnology, 103*(15), 6007–6021.

Shukla, R., Sarim, K. M., & Singh, D. P., (2020). Microbe-mediated management of arsenic contamination: Current status and future prospects. *Environmental Sustainability, 3*(1), 83–90.

Srivastava, S., Upadhyay, M. K., Tripathi, R. D., & Dhankher, O. P., (2016b). Arsenic transport, metabolism and toxicity in plants. *Int. J. Plant Environ., 2*, 1, 2. doi: 10.18811/ijpen.v2i1-2.6614.

Verma, S., Verma, P. K., & Chakrabarty, D., (2019). Arsenic bio-volatilization by engineered yeast promotes rice growth and reduces arsenic accumulation in grains. *Int. J. Environ. Res., 13*, 475–485. https://doi.org/10.1007/s41742-019-00188-7.

Webb, C., Campbell, C., Davis, A. D., Dawadi, S., Dixon, D., & Sorensen, J., (2006). Arsenic remediation of drinking water using limestone-based material. In: *Abstracts of Papers of the American Chemical Society* (Vol. 231). 1155 16th ST, NW, Washington, DC 20036 USA: Amer Chemical Soc.

World Health Organization [WHO], (2004). *Guidelines for Drinking-Water Quality*. Geneva: World Health Organization. [Internet]. Available at: https://www.who.int/publications/i/item/9789241549950 (accessed on 29 March 2023).

Wu, Z., Ren, H., McGrath, S. P., Wu, P., & Zhao, F. J., (2011). Investigating the contribution of the phosphate transport pathway to arsenic accumulation in rice. *Plant Physiol., 157*, 498–508. doi: 10.1104/pp.111.178921.

Yang, H. C., & Rosen, B. P., (2016). New mechanisms of bacterial arsenic resistance. *Biomed. J., 39*, 5–13. doi: 10.1016/j.bj.2015.08.003.

Zhang, S. Y., Williams, P. N., Luo, J., & Zhu, Y. G., (2017). Microbial mediated arsenic biotransformation in wetlands. *Frontiers of Environmental Science & Engineering, 11*(1), 1–11.

Zhao, F. J., Ma, J. F., Meharg, A. A., & McGrath, S. P., (2009). Arsenic uptake and metabolism in plants. *New Phytol., 181*, 777–794. doi: 10.1111/j.1469-8137.2008.02716.

CHAPTER 10

The Determination and Speciation of Arsenic in Rice

DIPAK KUMAR MAHIDA,[1] ANKITA PATEL,[1] and RAKESH RAWAL[2]

[1]*Department of Biochemistry and Forensic Science, Gujarat University, India*

[2]*Department of Life Science, Gujarat University, India*

ABSTRACT

Arsenic is a highly toxic and carcinogenic metalloid that is found in soil and contaminated water. It accumulates in rice when contaminated water is exposed to paddy fields and creates a major threat to human health. The anaerobic growing condition of flooded rice paddies and the physiological specificities of rice are major factors responsible for increased arsenic levels in rice. Periodically, different techniques are developed for the determination, separation, and quantification of arsenic in various parts of the rice plant. In this chapter, we discussed different determination techniques for arsenic and their speciation as well as extraction and separation methods for different parts of rice. The formation/transformation and chemical complexity of arsenic species (Arsenite, Arsenate, DMA, MMA, AsC, AsB, DMTAV, TMAO, Me^4As^+, DPAA, PAA, MPAA, DMPAO, MDPAO) is also discussed here. Inductively coupled plasma mass spectrometry (ICP-MS) and ion chromatography are the most common methods used for simultaneous speciation of arsenic. ICP-MS hyphenated to high-performance liquid chromatography (HPLC) or laser ablation (LA) is broadly used. Hydride generation atomic absorption spectrophotometry (HG-AAS) and DNA-gated graphene field effect transistors are specific non-chromatographic methods developed

Arsenic in Rice: A Global Menace. Garima Awasthi, Sudhakar Shrivastava, and Mahipal Singh Sankhla, (Eds.)

for As-III in rice. Flow Injection Hydride Generation Atomic Absorption Spectrometry (FI-HG-AAS) procedures are based on the *in-situ* formation of quaternary ammonium salt as a micro-sorbent. Inorganic speciation of arsenic is more harmful than organic arsenic in human and animal bodies. Fertilizers and microbial (mostly bacteria) impact on speciation of arsenic in rice plants. Arsenite (As-III), Arsenate (As-V), Monomethylarsonic acid (MMA), and dimethylarsinic acid (DMA) are the main arsenic species found in rice. Other species are present in more or less amounts according to environmental conditions. In the future, there might be more toxic and harmful species that will exist due to the polluted environment.

10.2 INTRODUCTION

Arsenic is one of the most important toxic and carcinogenic elements in the environment due to an increased risk to human health. Arsenic can be derived from nature or human activities (Costa et al., 2016). In the modern world, industrial growth and mining play an important role in the tremendous increase of arsenic levels in the environment. Soil, rocks, and water also contain a small amount of arsenic (Fang et al., 2016). When contaminated water comes in contact with fields, it is absorbed by the roots and contaminates plants.

Rice (*Oryza sativa*) is produced and consumed on a large scale in the world. Nowadays, it is becoming the main agricultural product with a high concentration of arsenic in food (Son et al., 2019). In a plant, the level of arsenic in various parts is in the following manner: Root>Straw>Husk>Whole grains>Husked rice. The reported level of arsenic in the rice sample is summarized in Table 10.1 (Fang et al., 2016).

TABLE 10.1 Reported Arsenic Levels in Rice in Various Countries

Name of the Country	Reported Level of Arsenic in Rice (mg/kg)
Bangladesh	<0.01–2.05
China	0.31–0.70
India	0.03–0.044
Taiwan	<0.10–0.76
US	0.11–0.66
Vietnam	0.03–0.47
Italy	0.08–0.38
Spain	0.08–0.38

The determination of arsenic in rice samples became necessary as a precaution against disease es caused by it. Periodically, various techniques are evaluated for extraction, separation, and determination of arsenic in different rice samples such as rice grains, rice flour, rice straw, and rice milk. Quantitative analysis of arsenic is challenging in rice samples because of matrix interference. Rice has a complex matrix containing hydrocarbons, proteins, fat, and fibers. For precise analysis, it is necessary to develop a standard analytical technique (Son et al., 2019). Maintaining good efficiency and original integrity is critical during the extraction of various species of arsenic. Microwave-assisted extraction (MAE) and heat-assisted extraction are mostly used (Narukawa et al., 2014).

The separation of arsenic is most commonly done by chromatographic technique. Two kinds of chromatography namely ion-pair and ion-exchange are used (Wang et al., 2012). There is a wide variety of determination techniques. Mainly used techniques are discussed in this chapter. Inductively coupled plasma mass spectrometry (ICP-MS) and hyphenated ICP-MS with high-pressure liquid chromatography (HPLC) or laser ablation (LA) are familiar to rice samples. In other techniques, Hydride generation atomic absorption spectrophotometry (HG-AAS) is used separately and in hyphenated mode. Solid phase extraction HG-AAS, flow injection HG-AAS, and electrothermal AAS are hybrid modes of atomic absorption spectrometers. UV visible spectrometry and DNA graphene field effect transistors are rarely used for determination. Speciation of arsenic indicates the identification and quantification of individual species. Inorganic and organic species are found in various forms.

This chapter discusses various determination techniques including extraction and separation of arsenic from rice. The basic information on arsenic species and their chemical structures is also discussed here.

10.2 ANALYTICAL METHODS FOR ARSENIC IN RICE

10.2.1 EXTRACTION PROCEDURE

Prior to any chemical analysis of any sample, they are required to extract the desired component contained in it. The extraction of arsenic from rice is the most challenging task during arsenic speciation analysis. Species of arsenic in rice are extracted by using corresponding solvent systems before their identification and quantification. Trifluoroacetic acid (TFA), nitric acid (HNO_3), hydrogen peroxide (H_2O_2), methanol, ethanol, acetonitrile, and water are compatible solvents and are widely used for the extraction of arsenic from rice. Sometimes, species of arsenic convert to other species during the extraction, such as As-V reduced into As-III.

Before the extraction, rice samples need some pretreatment that can affect the quality of the analysis. Mainly grinding and sieving are done for the preparation of the running sample. The grinding process converts whole rice into powder form and sieving could differentiate particles according to their size. About 125 μm–500 μm could be used for sieving (Pizzaro et al., 2003). Besides, when the sample is liquid, the clean-up step is required for dispelling protein and salts in the matrix. C_{18} cartridge extraction, dilution, and freezing could be done for this procedure. When the sample is solid, digestion of tissue is necessary before solvent extraction (Benrander, 1999).

Generally, rice grains, rice flour, rice water, rice straw, rice milk, etc., are used as samples for analysis of arsenic speciation. Many different techniques were used for the extraction of various species of arsenic from rice till today.

10.2.2.1 MICROWAVE-ASSISTED EXTRACTION

A microwave-assisted extraction (MAE) is a mostly used technique for arsenic speciation in rice. Generally, it is used for the extraction of active components from medicinal plants in pharmaceutical sciences. These techniques required microwave radiations for heating solvents containing samples (Gomez et al., 2020). In MAE, polar solvents are used to absorb microwave radiation. Methanol and ethanol are appropriate solvents, generally used in this technique (Wang et al., 2016) (Figure 10.1).

10.2.2.2 ULTRASONIC-ASSISTED EXTRACTION

The application of ultrasound energy could be done to facilitate sample pretreatment during the analysis of arsenic. It is widely used in industry, medicinal science, analytical chemistry, and forensic science. The main advantages of this technique are not required high temperature, pressure,

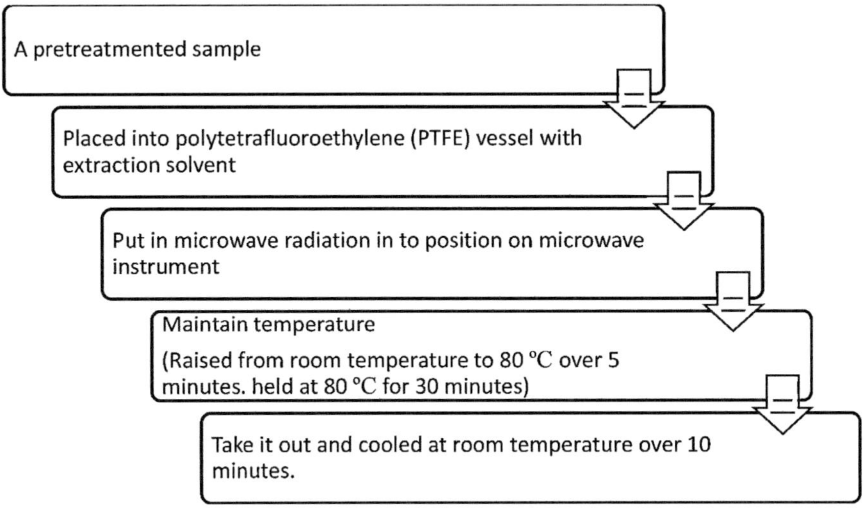

FIGURE 10.1 Microwave-assisted extraction procedure.

or concentrated and harmful chemicals. Besides, it is simple, easy, and efficient to use (Da La et al., 2013).

For ultrasonic-assisted extraction, samples have been placed into ultrasonic effects of acoustic cavitation. Which is produce vibrations and acceleration in solid and liquid particles which helps solutes to rapidly diffuse to solvents (Cares et al., 2009) (Figure 10.2).

10.2.2.3 MICROEXTRACTION

Microextraction techniques are becoming very popular in the past decade. These techniques are specially developed and improved for the extraction of analytes from a very small amount of sample. It could be based on microwave-assisted, ultrasonication, enzymatic, liquid-liquid extraction, or solid-phase extraction (Sarfraj et al., 2010).

A new approach to the extraction of analytes ultrasonic radiation and hydrolytic enzymes used combined which is known as Ultrasound assisted enzyme based hydrolytic water phase microextraction (UA-EH-WPME) method (Yilmaz et al., 2018) (Figure 10.3).

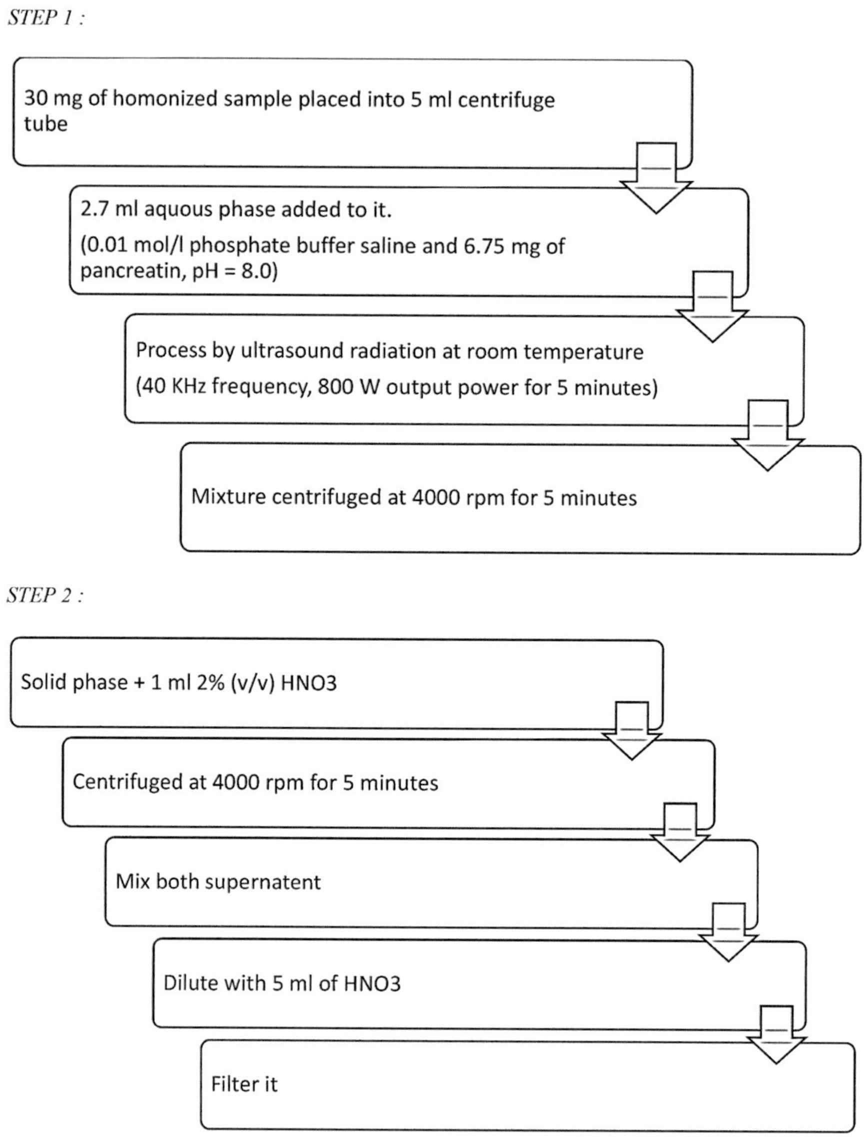

FIGURE 10.2 Ultrasound-assisted extraction procedure in two steps.

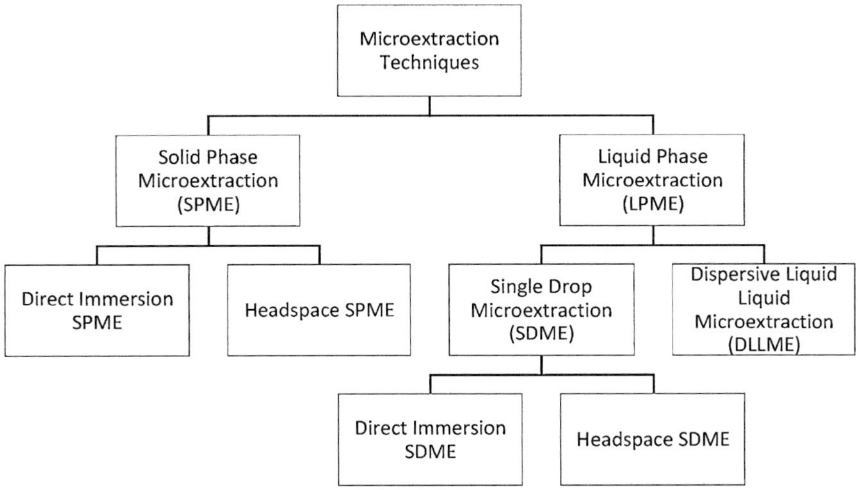

FIGURE 10.3 Classification of microextraction techniques.

10.2.2.4 ACCELERATED SOLVENT EXTRACTION

Accelerated solvent extraction also known as pressurized liquid extraction uses upraised temperature during extraction. This technique was used and concluded that ASE is inoperative for the extraction of arsenic from rice samples (Heitkemper et al., 2001). The reason behind the failure of ASE for arsenic extraction in rice is high temperatures and pressures give rise to the swelling of rice samples (Wang et al., 2012).

10.2.2.5 ACID LEACHING

Acid leaching is a classical extraction method done by using HNO_3 and trifluoroacetic acid (TFA). For the extraction procedure, an appropriate amount of rice powder is simply added to the 2 ml of 2 mol/L TFA and incubated at 100 C for 6 hours (Meharg et al., 2008). This technique is more efficient and higher-ranking compared to other extraction techniques (Heitkemper et al., 2001). Although, acid leaching has a disadvantage that creates problems during analysis: (i) TFA converts the arsenate into arsenite sometimes (Abedin et al., 2012); (ii) the species gives poor resolution against chromatographic plate during the separation when it was extracted by TFA (Pizzaro et al., 2003). HNO_3 with different ratios of concentration is also used to propose an extraction. About 0.28 mol/L HNO_3 at 95°C–100°C for 90 minutes is an optimal condition (Huang et al., 2010).

10.2.2.6 SOLVENT EXTRACTION

For extraction of arsenic speciation from biosamples, water is also a conventional and efficient solvent. Up to 97% of recovery has been reported in studies when water is used as a solvent in extraction (Narukawa et al., 2010). Around 1% HNO_3, chloroform, and diluted HCl can also be used for this method. Around 1% HNO_3 is more efficient for rice milk samples (Wang et al., 2012).

10.2.2.7 ENZYMATIC EXTRACTION

The enzymatic technique was developed by Glahn et al. This technique includes some conditional environments as in the human digestive system, more similar to the stomach and intestine. It was created for estimating the bioavailability of iron in food products (Ackerman et al., 2005).

10.2.2.8 OTHER TECHNIQUES

Nowadays, Cloud point extraction and Soxhlet extraction are also utilized in arsenical extractions from rice.

Here, are some important parameters to be taken into consideration during the extraction of arsenic species from rice:

- Washing of rice sample;
- Different liquid-to-solid ratios;
- Temperature;
- Hold time;
- Particle size.

Washing the rice before the extraction could affect the quality of the analysis. Contaminated water can be increasing the level of arsenical speciation in the sample or distilled water can decrease the level of arsenic in the sample (Williams et al., 2006). The relation between the efficiency of extraction and other parameters is shown below, also optimal stages of their parameters are reported in Table 10.2 (Narukawa et al., 2010).

TABLE 10.2 Relation between Extraction Efficiency and Various Parameters

Sl. No.	Parameter	Relation with Extraction Efficiency	Optimal Stage
1.	Liquid/solid ratio	? ∝ L/S ratio	10
2.	Temperature	? ∝ T	80°C
3.	Hold time	? ∝ Time	30 minutes
4.	Size of particles	? ∝ 1/size	–

10.2.3 SEPARATION PROCEDURES

The separation of molecules is important for reducing the limit of detection, better quantification, and confident identification. Chromatographic separation is the most precise technique due to the necessary selectivity and purity of the result. The separation of arsenic speciation from rice is most often done by liquid chromatography because of its accuracy, sensitivity, and precision (Wang et al., 2016). Liquid chromatography is divided as in Figure 10.4.

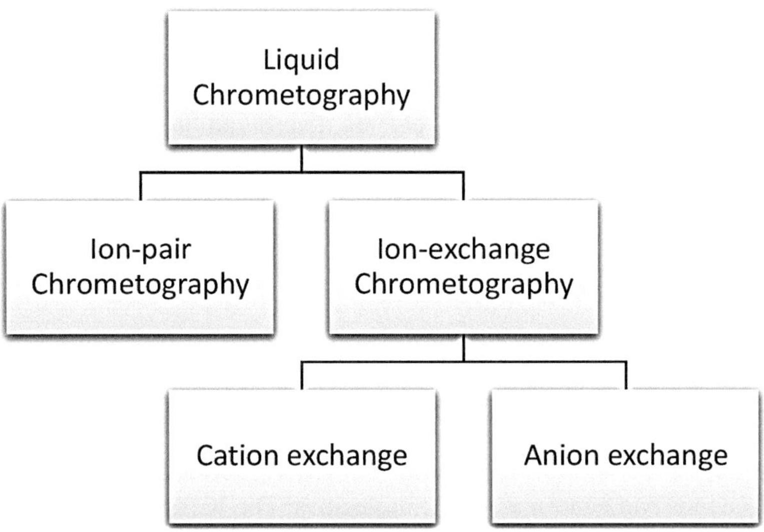

FIGURE 10.4 Classification of liquid chromatography.

Ion-pair chromatography and Ion-exchange chromatography both are helpful for the detection of arsenic species in rice. But ion-pair chromatography is not commonly used due to the complex matrix of rice. Ion exchange chromatography is used for the separation of ionizable charged molecules known as resins. Widely it is used for the separation of proteins, peptides, amino acids, and nucleotides. But in past decades, it is utilized in the separation of metallic elements such as arsenic, lead, and iron. Two types of ion exchange chromatography exist based on the charge of the molecule: anion exchanger and cation exchanger (Zamfir et al., 2021).

Ion exchange chromatography is mainly performed in four stages: (i) equilibration; (ii) loading; (iii) washing; and (iv) elution.

Different species are separated by using the most appropriate column and their conventional mobile phases. Many scientists have attempted different solvents for establishing the most accurate solvent system (Wang et al., 2016).

10.2.4 DETERMINATION PROCEDURES

Determination techniques are analytical methods that generate more readable data on chemical identity and quantity in a broad sense. Generally, two kinds of determination techniques are available: classical techniques and instrumental techniques. Classical determination has been done with wet chemical methods and instrumental techniques through various appropriate instruments. In rice, arsenic determination is mostly done by instruments due to the small quantities of arsenic in samples.

10.2.4.1 INDUCTIVELY COUPLED PLASMA MASS SPECTROMETRY (ICP MS)

Inductively coupled plasma mass spectrometry (ICP-MS) is a type of mass spectrometry that is capable of detecting metals at very low concentrations. ICP-MS combines a high-temperature source with a mass spectrometer. Plasma is one of the forms of states of matter; it consists of free electron positive ions and neutral atoms or molecules. The ICP source converts the atoms of the elements into ions in a sample. These ions are then separated and detected by mass spectrometry (Montaser et al., 1998). ICP is a plasma that is ionized by inductively heating the gas with an electromagnetic coil. Argon gas is electrically heated to a high-temperature plasma. The ion enters into an electric field and is separated according to its mass/charge (m/z) ratio. The signal intensities are directly proportional to the concentrations of the elements in the sample (Forsgard et al., 2007).

An ICP-MS becomes a standard method due to its high specificity and selectivity. ICP-MS can be determining the total arsenic speciation in acid-digested rice samples using 1.0% (v/v) nitric acid and multi-element determination can be performed (Jinadasa et al., 2020). A main issue in the utilization of ICP-MS for arsenic analysis in rice is the formation of the polyatomic species 40Ar35Cl, which can notably obstruct the analysis of arsenic in rice (Wang et al., 2012).

10.2.4.1.1 High-Pressure Liquid Chromatography – Inductively Coupled Plasma Mass Spectrometry (HPLC ICP MS)

The quantification of arsenic in rice samples can be done by using HPLC-ICP-MS. It can be performed by using the standard addition technique (Jinadasa et al., 2020). The standard addition technique is used when the extract contains multi-constituents that absorb or influence the absorbance of the analyte. The method is effective in compensating for spectral and chemical interferences caused by the sample matrix. It implies the addition of one or more increments of standard solution to the analyte sample solution of the same size. Diluting the solutions to a fixed total volume Aspirating various solutions into flame and measuring their absorbance (Garcia, 2012).

10.2.4.1.2 Laser Ablation Inductively Coupled Plasma Mass Spectrometry (LA ICP MS)

LA ICP MS is a highly sensitive instrumental method for isotopic analysis, it is more compatible for solid samples. Laser ablation is the procedure of focusing a laser beam on a sample surface to generate fine particles. The ablated particles were transited to the hyphenated ICP MS for the digestion and ionization of the samples. Mass spectrometer detectors catch the excited ions and subsequently give analyzed data (Russo et al., 1999). For fs LA ICP MS, cross-sectioned rice can be run directly without any other preparation at an instrumental condition of 5.76 mm^2 ablated area with a 40×40 grid point (Choi et al., 2014) (Figure 10.5).

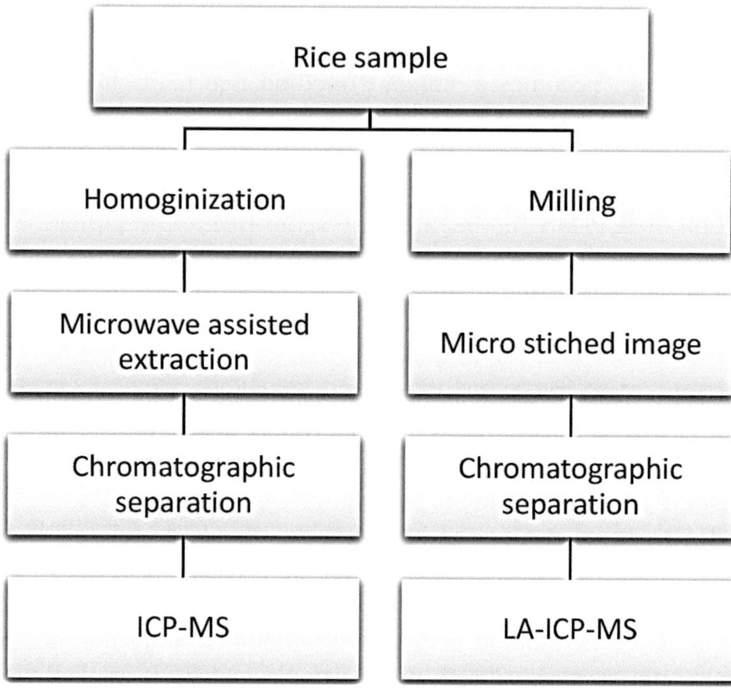

FIGURE 10.5 More familiar sample preparation methods for ICP-MS and LA-ICP-MS.

10.2.3.2 HYDRIDE GENERATION ATOMIC ABSORPTION SPECTROPHOTOMETRY (HG AAS)

In 1950, Alan Walsh developed a technique that can give quantitative data of about 70 elements including metallic elements similar to aluminum, cadmium, arsenic, lead, iron, zinc, etc., known to be absorption mass spectrometry (AAS). AAS is particularly useful for determining trace essence in liquid and is mostly independent of the molecular form of the essence in the sample (Welz et al., 2008).

Hydride generation atomic absorption spectroscopy (HG-AAS) is available via an option for multitudinous modern AAS instruments. It only needs the hydride generation module. Nowadays, HGAAS is routinely used for the analysis of metalloids like antimony, arsenic, selenium, and tellurium. Multitudinous corridors of the HGAAS system are identical to that of AAS a hollow cathode lamp, air/acetylene atomizer, and detector, and the fairly complex hydride generation system. The nebulizer demanded in AAS is not used in HGAAS. The system described is also a continuous flux system, but batch flux systems have been used throughout history (Ceirveira et al., 2015).

The response of multitudinous metalloid oxyanions with sodium borohydride and HCl produces a changeable hydride. The oxidation state of the arsenic is vital with AAS, and care must be taken to produce the specific metalloid oxidation state before the sample is introduced into the hydride generation system. The other important point is the time from reagent mixing and when the changeable hydride is isolated from the liquid and transferred to the optical cell. The timing of that process is controlled by flowing reagents together using a peristaltic pump and tubing of specific lengths. After being mixed the liquid amalgamation flows through a tube of a specific length (read this as a controlled response time) and is ultimately flowed into a gas/liquid division where the hydride and some gaseous hydrogen (produced by the NaBH4+H2 response) bubble out and are refined (via a high chastity inert gas) into the optical cell via a gas shifting line. Most of the reagents introduced into the system flow to a waste vessel, and since the acid content is truly high, constantly approaching 50, as with AAS, the waste vessel is glass and must be handled precisely and labeled well (Denina et al., 1995).

10.2.4.2.1 Solid Phase Extraction – Hydride Generation Atomic Absorption Spectrophotometry (SPE HG AAS)

This approach was developed for the analysis of seafood samples, but nowadays it is used for the determination of inorganic arsenic (Asi) rice samples. For this technique, arsenic is separated by using silica-based ion exchange solid phase extraction (SPE) followed by loading of the buffered sample extract on the SPE column and washed with 0.5 M acetic acid (Rasmussen et al., 2013). The maintenance of pH in the range of 5.0–7.5 is necessary for this procedure which can be controlled by pre-conditioning the cartridge with 2 ml methanol and balancing it with the buffer solution. Rasmussen et al. compared the results of analyzed samples of marine origin by SPE-HG-AAS with the sets of results of the same sample analyzed by HPLC-ICP-MS and concluded that the results were not notably different (Rasmussen et al., 2012).

10.2.4.2.2 Flow Injection – Hydride Generation Atomic Absorption Spectrophotometry (FI HG AAS)

FI HG AAS is a rapid, simple, accurate, and precise method for analyzing environmental and biological standard reference materials. Acid digestion and dry aching technique can destroy the organic components in biological samples (Samanta et al., 1999). The *in-situ* formation of quaternary ammonium salts is performed an important role in this procedure. Tetra-n-butylammonium bromide (TBAB) is injected in a small amount into the sample solution as a phase-transfer agent. Also, a small portion of perchlorate is added after shaking well. This perchlorate works as an ion-pairing agent which interacts with the sample solution and forms fine solid particles. This procedure gives accuracy in results in a short extraction time with good dispersion and great recovery (Dos Santos et al., 2021).

10.2.4.2.3 Electrothermal Atomic Absorption Spectrophotometry (ETAAS)

ETAAS is based on the heating of graphite tubes by an electric charge which converts graphite tubes into atom cells. This heating mechanism increases the accuracy of results. The tube temperature can have differed temperatures from sample to sample and from standard to sample which can be controlled by a computer system. A small amount of sample (generally 20–50 μL) in a dilute aqueous solution form is introduced. This technique creates a problem when the analyte is highly volatile or when the analyte and matrix both are volatile at similar temperatures (Edbon et al., 1998).

In literature, these techniques are notably applied for the determination of some compounds such as magnesium and palladium in biological samples. Also, it is utilized for the determination of arsenic in whole rice and rice flour (Dos Santos et al., 2015).

10.2.4.3 UV-VIS SPECTROMETRY

Ultraviolet (UV)-Visible spectrometry is widely used for drug identification, metal detection, bacteria culturing, and quantification of compounds.

This analytical technique is based on measurements of discontinued wavelengths of UV or visible lights that are soaked up or transmitted by samples. The energy of light is inversely proportional to its wavelength, which is useful to analyze or identify various components. A certain amount of energy encourages the electrons into samples to a higher energy state which is detectable as absorption (Perkampus et al., 2013). For the determination of arsenic in rice samples, a UV visible spectrometer was used in a limited number of studies due to not having any extra special features regarding metal detection from food samples.

10.2.4.4 DNA GRAPHENE FIELD EFFECT TRANSISTORS

DNA graphene field effect transistors are newly developed and rarely used techniques for specific, fast, and precise determination of As-III. In this technique, the specific interaction between As-III and gold disrupted the adsorption states of DNA probes, redistributing surface charges on the gate electrode, further leading to potential drop changes at the interfaces of the gate electrode and graphene active layer (Wang et al., 2016).

10.3 SPECIATION OF ARSENIC IN RICE

Identification and quantification of individual arsenic species perform an important role to recognize the distribution of toxicity and metabolism of arsenic. Arsenic is distributed in the form of more than 100 compounds in the environment and biological systems. Their toxicity is directly based on chemical structures and oxidation states (Reid et al., 2020). Fertilizers and microorganisms can affect the speciation of arsenic in rice plants, leading to the accumulation of both organic and inorganic arsenic species. For example, the use of arsenic-containing fertilizers can increase the uptake of arsenic by rice plants, resulting in higher levels of arsenic in the rice grains. Similarly, some microorganisms, such as bacteria, can transform inorganic arsenic species into organic arsenic species. In rice, the main arsenic species found are arsenite (As-III), arsenate (As-V), monomethyl arsenic acid (MMA), and dimethylarsinic acid (DMA). However, other arsenic species may also be present in varying amounts depending on environmental conditions. In a study of speciation, mainly three factors are taken into consideration:

1. **Stability:** It is a major concern for the development of detection techniques for arsenic in rice.
2. **Formation/Transformation:** It is also important to recognize the entry of arsenic from where it comes to the plant (From soil/water/bio-organism to rice plants).
3. **Chemical Complexity:** The structure and Behavior of arsenic species help to diagnose symptoms and effects of species on plants and animals.

Table 10.3 contains the list of all arsenic species founds in a rice plant with their chemical formula and molecular weight (Wang et al., 2012).

TABLE 10.3 List of Various Arsenic Species Founds in Rice Samples

SL. No.	Name of Arsenic Species	Chemical Formula	Molecular Weight (Da)
1.	Arsenite (As-III)	AsO_3^{-3}	165.09
2.	Arsenate (As-V)	AsO_4^{-3}	138.919
3.	dimethylarsinic acid (DMA)	$C_2H_7AsO_2$	137.9977
4.	Monomethylarsonic acid (MMA)	CH_5AsO_3	139.970
5.	Arsenocholine (AsC)	$C_5H_{14}AsO^+$	165.09
6.	Arsenobetaine (AsB)	$C_5H_{11}AsO_2$	177.9975
7.	Dimethylthioarsinic acid (DMTAV)	C_2H_7AsOS	154.07
8.	Trimethylarsine oxide (TMAO)	C_3H_9AsO	136.02
9.	Tetramethylarsonium (Me_4As^+)	$C_4H_{12}AsI$	261.96
10.	Diphenylarsinic acid (DPAA)	$C_{12}H_{11}AsO_4$	294.13
11.	Phenylarsonic acid (PAA)	$C_8H_8AsO_2$	136.15
12.	Methylphenylarsinic acid (MPAA)	$C_7H_9AsO_2$	200.07
13.	Dimethylphenylarsine oxide (DMPAO)	$C_8H_{11}AsO$	198.09
14.	Methyldiphenylarsine oxide (MDPAO)	$C_{13}H_{13}AsO$	260.16

10.3.1 ARSENITE (AS-III)

Arsenite is the most common and highly toxic species of arsenic in rice samples. It is well known as As-III due to its +3-oxidation state. In a field, mainly it is distributed in the groundwater where it contaminates rice plants. As-III is a more mobile species than other arsenic species, it is quite problematic during determination and identification. In rice, commonly it is found as oxides of iron. Such as ferrihydrite significantly affects dissolved arsenic concentration and limits the mobility of As(III) and As(V) (Figure 10.6).

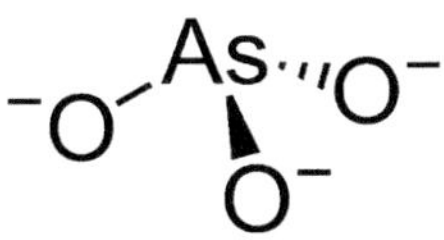

FIGURE 10.6 Structure of arsenite (As-III).

10.3.2 ARSENATE (AS-V)

Arsenates are salts or esters of arsenic acid. It has a valency of 5 as it is known as a pentavalic arsenic (As-V). Arsenate and phosphates allow similar pathways in a higher plant. Its primary toxicity is based on the inhibition of proteins that contain phosphates. In competitive conditions, As-III converts into As-V which is considered as a limitation of some extraction techniques (Figure 10.7).

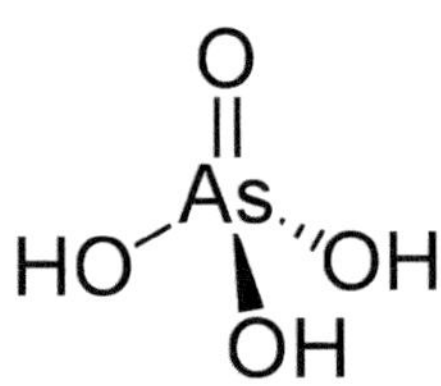

FIGURE 10.7 Structure of arsenate (As-V).

10.3.3 MONOMETHYLARSONIC ACID (MMA) AND DIMETHYLARSINIC ACID (DMA)

Sometimes, Asi is converted by biological systems into organic arsenic species commonly methylated species such as monomethyl arsenic acid (MMA) or dimethyl arsenic acid (DMA). DMA is highly toxic than the Asi species, generally placed into plants. In rice, DMA can cause straighthead disease due to its neutral pKa value. DMA presents in two forms:

dissociated (anion) or undissociated (cation) forms in paddy field soil and rice plants (Figures 10.8 and 10.9).

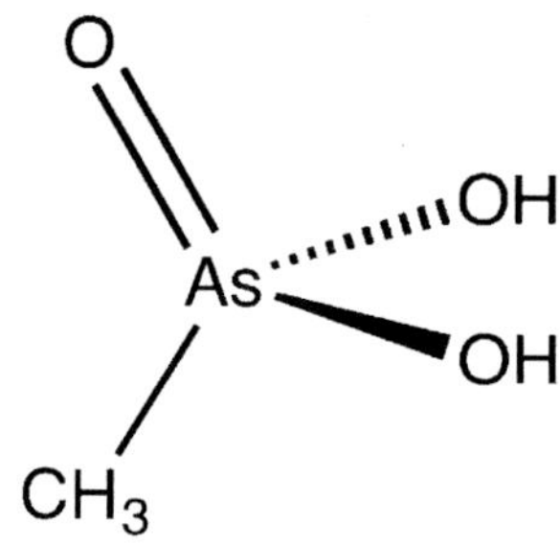

FIGURE 10.8 Structure of monomethyl arsenic acid (MMA).

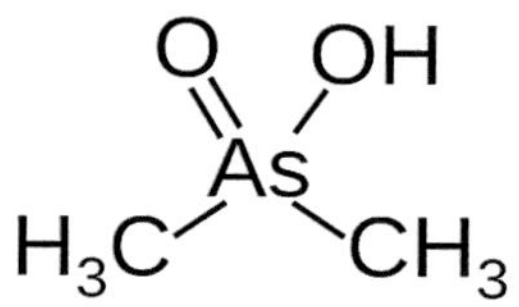

FIGURE 10.9 Structure of dimethyl arsenic acid (DMA).

10.3.4 OTHER SPECIES

Arsenocholine (AsC), arsenobetaine (AsB), dimethylthioarsinic acid (DMTAV), trimethylarsine oxide (TMAO), tetramethylarsonium (Me_4As^+), diphenylarsinic acid (DPAA), phenylarsonic acid (PAA), methylphenylarsinic acid (MPAA), dimethylphenylarsine oxide (DMPAO), methyldiphenylarsine oxide (MDPAO) are also founded in a more or less amounts in rice plants (Figures 10.10–10.14).

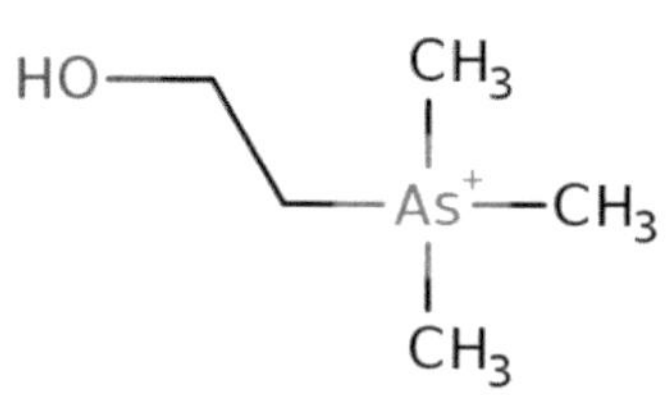

FIGURE 10.10 Structure of arsenocholine (AsC).

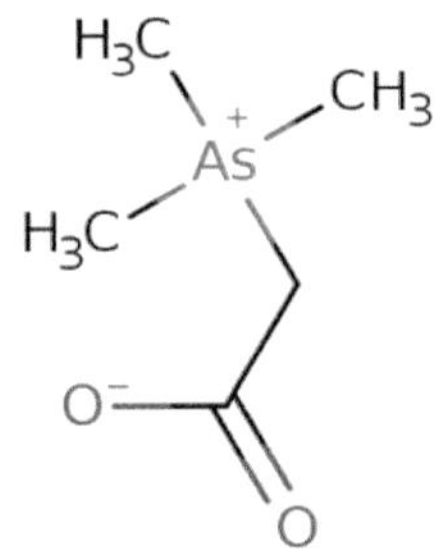

FIGURE 10.11 Structure of arsenobetaine (AsB).

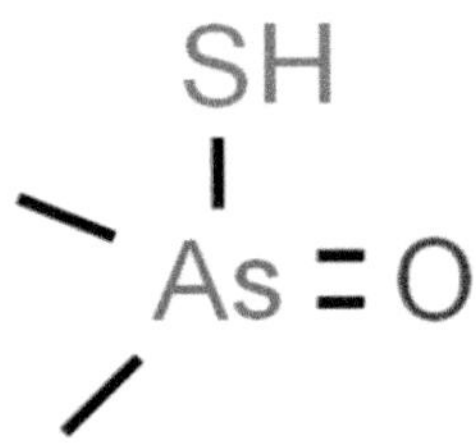

FIGURE 10.12 Structure of dimethylthioarsinic acid (DMTAV).

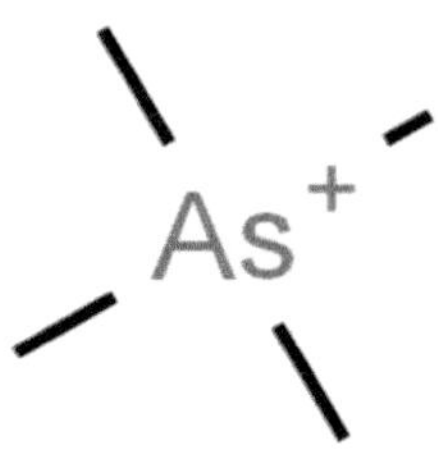

FIGURE 10.13 Structure of tetramethylarsonium.

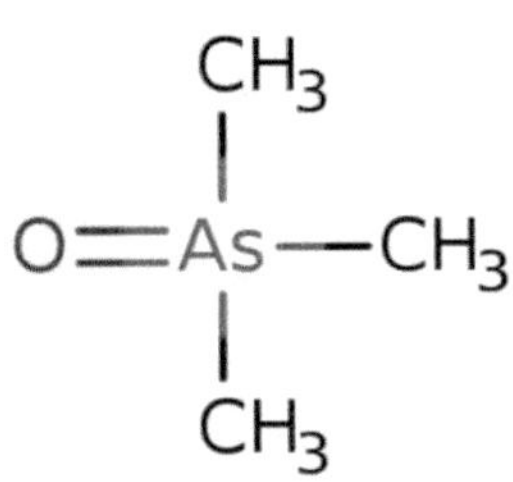

FIGURE 10.14 Structure of trimethylarsine oxide (TMAO).

10.4 CONCLUSION

An accumulation of arsenic in rice from soil, rocks, and water can be determined by various instrumental analytical techniques. These techniques are evaluated over time. Different analytical methods for the determination and speciation of arsenic in rice are discussed in the present chapter. Microwave-assisted extraction is a reliable, more accurate technique for any kind of rice sample. Liquid/solid ratio, temperature, hold time and size of sample particles are the most important parameters with respect to extraction efficiency. This chapter does not include a comprehensive review but with respect to the analysis of arsenic in rice, the analytical methods demanded more advanced techniques which can provide more detailed information of the present arsenic species. Instruments need to be intelligent to understand the complexity of food matrices and chemical structures of arsenic species. They also need to be able to identify all the species by one instrument.

The amount of arsenic in food products keeps increasing due to continuous industrial growth and increasing pollution. In the future, there is a possibility that determination of arsenic in rice will be necessary before its consumption as food and there might be more injurious species that will be in existence.

KEYWORDS

- **arsenic**
- **determination**
- **extraction**
- **microwave-assisted extraction**
- **rice**
- **speciation**
- **tetramethylarsonium**
- **trimethylarsine oxide**

REFERENCES

Abedin Md, J., et al., (2002). Arsenic accumulation and metabolism in rice (*Oryza sativa* L.). *Environmental Science & Technology, 36*(5), 962–968.

Ackerman, A. H., et al., (2005). Comparison of a chemical and enzymatic extraction of arsenic from rice and an assessment of the arsenic absorption from contaminated water by cooked rice. *Environmental Science & Technology, 39*(14), 5241–5246.

Benramdane, L., Bressolle, F., & Vallon, J. J., (1999). Arsenic speciation in humans and food products: A review. *Journal of Chromatographic Science, 37*(9), 330–344.

Cares, M. G., et al., (2010). Ultrasonically assisted extraction of bioactive principles from *Quillaja Saponaria* Molina. *Physics Procedia, 3*(1), 169–178.

Cerveira, C., et al., (2015). Speciation of inorganic arsenic in rice using hydride generation atomic absorption spectrometry (HG-AAS). *Analytical Methods, 7*(11), 4528–4534.

Choi, S. H., et al., (2014). Analysis of arsenic in rice grains using ICP-MS and fs LA-ICP-MS. *Journal of Analytical Atomic Spectrometry, 29*(7), 1233–1237.

Costa, B. E. S., et al., (2016). Analytical strategies for the determination of arsenic in rice. *Journal of Chemistry, 2016*.

De La Calle, I., et al., (2013). Ultrasound-assisted single extraction tests for rapid assessment of metal extractability from soils by total reflection X-ray fluorescence. *Journal of Hazardous Materials, 260*, 202–209.

Dedina, J., & Tsalev, D. L., (1995). *Hydride Generation Atomic Absorption Spectrometry* (Vol. 130, p. 526). Wiley.

Dos, S. C., Bruno, E., & Nívia, M. M. C., (2021). Selective determination of As (III) and total inorganic arsenic in rice sample using *in-situ* μ-sorbent formation solid phase extraction and FI-HG AAS. *Journal of Food Composition and Analysis, 95*, 103686.

Dos, S. C., Bruno, E., Nívia, M. M. C., & Luciana, M. C., (2015). Determination of arsenic species in rice samples using CPE and ETAAS. *Food Chemistry, 178*, 89–95.

Ebdon, L., Andy, S. F., & Hill, S. J., (1998). *An Introduction to Analytical Atomic Spectrometry*. John Wiley & Sons.

Fang, Y., et al., (2016). Simultaneous determination of arsenic and mercury species in rice by ion-pairing reversed phase chromatography with inductively coupled plasma mass spectrometry. *Food Chemistry, 213*, 609–615.

Forsgard, N., (2007). *Inductively Coupled Plasma Spectrometry for Speciation Analysis: Development and Applications*. Diss. Acta Universitatis Upsaliensis.

García, R., & Báez, A. P., (2012). Atomic absorption spectrometry (AAS). *Atomic Absorption Spectroscopy, 1*, 1–13.

Gomez, L., Brijesh, T., & Garcia-Vaquero, M., (2020). Emerging extraction techniques: Microwave-assisted extraction. *Sustainable Seaweed Technologies* (pp. 207–224). Elsevier.

Heinz-Helmut, P., (2013). *UV-VIS Spectroscopy and its Applications*. Springer Science & Business Media.

Heitkemper, D. T., et al., (2001). Determination of total and speciated arsenic in rice by ion chromatography and inductively coupled plasma mass spectrometry. *Journal of Analytical Atomic Spectrometry, 16*(4), 299–306.

Jen-How, H., Gunter, I., & Peter, F., (2010). Quantitative chemical extraction for arsenic speciation in rice grains. *Journal of Analytical Atomic Spectrometry, 25*(6), 800–802.

Jinadasa, K. K., et al., (2020). Ionic imprinted polymer–vortex-assisted dispersive micro-solid phase extraction for inorganic arsenic speciation in rice by HPLC-ICP-MS. *Talanta, 220*, 121418.

Meharg, A. A., et al., (2008). Inorganic arsenic levels in rice milk exceed EU and US drinking water standards. *Journal of Environmental Monitoring, 10*(4), 428–431.

Montaser, A., (1998). *Inductively Coupled Plasma Mass Spectrometry*. John Wiley & Sons.

Narukawa, T., & Koichi, C., (2010). Heat-assisted aqueous extraction of rice flour for arsenic speciation analysis. *Journal of Agricultural and Food Chemistry, 58*(14), 8183–8188.

Pizarro, I., et al., (2003). Evaluation of stability of arsenic species in rice. *Analytical and Bioanalytical Chemistry, 376*(1), 102–109.

Rasmussen, R. R., et al., (2012). Development and validation of an SPE HG-AAS method for determination of inorganic arsenic in samples of marine origin. *Analytical and Bioanalytical Chemistry, 403*(10), 2825–2834.

Rasmussen, R. R., Yiting, Q., & Jens, J. S., (2013). SPE HG-AAS method for the determination of inorganic arsenic in rice—Results from method validation studies and a survey on rice products. *Analytical and Bioanalytical Chemistry, 405*(24), 7851–7857.

Reid, M. S., et al., (2020). Arsenic speciation analysis: A review with an emphasis on chromatographic separations. *TrAC Trends in Analytical Chemistry, 123*, 115770.

Russo, R. E., et al., (1999). Time-resolved plasma diagnostics and mass removal during single-pulse laser ablation. *Applied Physics A, 69*(1), S887–S894.

Samanta, G., et al., (1999). Flow injection hydride generation atomic absorption spectrometry for determination of arsenic in water and biological samples from arsenic-affected districts of West Bengal, India, and Bangladesh. *Microchemical Journal, 62*(1), 174–191.

Sarafraz-Yazdi, A., & Amirhassan, A., (2010). Liquid-phase microextraction. *TrAC Trends in Analytical Chemistry, 29*(1), 1–14.

Son, S. H., et al., (2019). An alternative analytical method for determining arsenic species in rice by using ion chromatography and inductively coupled plasma-mass spectrometry. *Food Chemistry, 270*, 353–358.

Wang, H., Jie, D., & Nanqi, R., (2016). Recent advances in microwave-assisted extraction of trace organic pollutants from food and environmental samples. *TrAC Trends in Analytical Chemistry, 75*, 197–208.

Wang, Z., & Don, F., (2006). Methods for the determination of arsenic speciation in rice: A review. *Encyclopedia of Analytical Chemistry: Applications, Theory and Instrumentation.*

Welz, B., & Michael, S., (2008). *Atomic Absorption Spectrometry*. John Wiley & Sons.

Yilmaz, E., (2018). Use of hydrolytic enzymes as green and effective extraction agents for ultrasound assisted-enzyme based hydrolytic water phase microextraction of arsenic in food samples. *Talanta, 189*, 302–307.

Zamfir, A. D., Raluca, I., & Mirela, S., (2021). Liquid-phase separation methods hyphenated to electrospray ionization mass spectrometry for structural analysis of chondroitin/dermatan sulfate. *Carbohydrate Analysis by Modern Liquid Phase Separation Techniques*, 529–562.

Index

B

D

E

F

G

J

K

L

M

P

T

U

V

W

X

Z

For Product Safety Concerns and Information please contact our
EU representative GPSR@taylorandfrancis.com Taylor & Francis
Verlag GmbH, Kaufingerstraße 24, 80331 München, Germany